Student Worksheets

Visual Anatomy & Physiology

Frederic H. Martini, Ph.D.
University of Hawaii at Manoa

William C. Ober, M.D.
Washington and Lee University

Judi L. Nath, Ph.D.
Lourdes College

Edwin F. Bartholomew, M.S.
Contributing Author

Claire W. Garrison, R.N.
Medical & Scientific Illustration
Contributing Illustrator

Kathleen Welch, M.D.
Fellow, American Academy of Family Practice
Clinical Consultant

Benjamin Cummings

Boston   Columbus   Indianapolis   New York   San Francisco   Upper Saddle River
Amsterdam   Cape Town   Dubai   London   Madrid   Milan   Munich   Paris   Montréal   Toronto
Delhi   Mexico City   São Paulo   Sydney   Hong Kong   Seoul   Singapore   Taipei   Tokyo

Executive Editor: *Leslie Berriman*
Project Editor: *Robin Pille*
Editorial Development Manager: *Barbara Yien*
Development Editor: *Alan Titche*
Associate Editors: *Katie Seibel and Kelly Reed*
Editorial Assistant: *Nicole McFadden*
Senior Managing Editor: *Deborah Cogan*
Senior Production Project Manager: *Nancy Tabor*
Production Management and Composition: *S4Carlisle Publishing Services, Inc.*

Design Manager: *Mark Ong*
Interior Designer: *Gibson Design Associates*
Main Text Cover Designer: *Jana Anderson*
Supplement Cover Designer: *Seventeenth Street Studios*
Senior Photo Editor: *Donna Kalal*
Photo Researcher: *Maureen Spuhler*
Senior Manufacturing Buyer: *Stacey Weinberger*
Market Development Manager: *Brooke Suchomel*
Marketing Manager: *Derek Perrigo*

Cover Photo Credit: © Tim Tadder / CORBIS All Rights Reserved.

Credits and acknowledgments for materials borrowed from other sources and reproduced, with permission, in this textbook appear after the Answers.

Notice: Our knowledge in clinical sciences is constantly changing. The authors and the publisher of this volume have taken care that the information contained herein is accurate and compatible with the standards generally accepted at the time of the publication. Nevertheless, it is difficult to ensure that all information given is entirely accurate for all circumstances. The authors and the publisher disclaim any liability, loss, or damage incurred as a consequence, directly or indirectly, of the use and application of any of the contents of this volume.

ISBN 10: 0-321-76827-2
ISBN 13: 978-0-321-76827-8
ISBN 10: 0-321-74199-4 (ValuePack Only)
ISBN 13: 978-0-321-74199-8 (ValuePack Only)

Benjamin Cummings
is an imprint of

www.pearsonhighered.com

2 3 4 5 6 7 8 9 10—CRW—15 14 13 12 11

Contents

Visual Anatomy & Physiology is designed with frequent opportunities to pause and practice, helping you to pace your learning throughout the chapter. This workbook gives you a space apart from your book to complete the Section Reviews and to practice with the Visual Outline with Key Terms.

Each of the **Section Review** pages from the textbook is reproduced twice in this workbook, so you can work these exercises more than once—without writing in your textbook. Complete the Section Review once after reading the chapter, and again right before your exam to help you review the material. Answers to the Section Reviews are included in the back of this booklet.

The **Visual Outline with Key Terms** from the end of each chapter in your textbook is also included in this workbook. Each module section from the outline is reproduced with space to write. Summarize the content of the module in your own words, using the terms in the order provided. If you can't remember a term, look it up in that module. Terms marked with a darker bullet are boldfaced in that module.

Don't forget: Lots of additional practice is available in the Study Area of **MasteringA&P**™ at www.masteringaandp.com. Practice Anatomy Lab™ (PAL™), A&P Flix™ animations, MP3 Tutor Sessions, Interactive Physiology®, and PhysioEx™ are all included to help you succeed in your A&P course.

1. Vocabulary

For each of the following descriptions, write the characteristic of living things described in the corresponding blank.

a   Usually refers to the absorption and utilization of oxygen and the generation and release of carbon dioxide

b   Indications that an organism is successful

c   Changes in the behavior, capabilities, or structure of an organism

d   Movement of fluid within the body; may involve a pump and a network of special vessels

e   The elimination of chemical waste products generated by the body

f   The chemical breakdown of complex structures for absorption and use by the body

g   Transports materials around the body of a large organism; changes orientation or position of a plant or immobile animal; moves mobile animals around the environment (locomotion)

h   Indicates that the organism recognizes changes in the internal or external environment

a   _respiration_

b   _reproduction_

c   _adaptability_

d   _circulation_

e   _elimenation_

f   _digestion_

g   _movement_

h   _responsiveness_

2. Matching

Write each of the following terms under the proper heading.

- Right atrium
- Myocardium
- Valve to aorta opens
- Left ventricle
- Valve between left atrium and left ventricle closes
- Pressure in left atrium
- Electrocardiogram
- Endocardium
- Superior vena cava

Anatomy

right atrium
myocardium
left ventricle
endocardium
superior vena cava

Physiology

Valve to aorta opens
Valve btn L+R Vent. closes
pressure in L atrium
electrocardiogram

3. Short answer

Briefly describe how the relationship of form and function of a house key and its front door lock are both similar to and different from a chemical messenger and its receptor protein.

Only certain chem. messengers work with particular receptor proteins. Just like a lock + key.

4. Section integration

How might a large organism's survival be affected by an inadequate internal circulation network?

W/o adequate internal circulation cells would begin to die thus causing damage to organs + organ systems. eventually death.

1. Vocabulary

For each of the following descriptions, write the characteristic of living things described in the corresponding blank.

a   Usually refers to the absorption and utilization of oxygen and the generation and release of carbon dioxide

b   Indications that an organism is successful

c   Changes in the behavior, capabilities, or structure of an organism

d   Movement of fluid within the body; may involve a pump and a network of special vessels

e   The elimination of chemical waste products generated by the body

f   The chemical breakdown of complex structures for absorption and use by the body

g   Transports materials around the body of a large organism; changes orientation or position of a plant or immobile animal; moves mobile animals around the environment (locomotion)

h   Indicates that the organism recognizes changes in the internal or external environment

a _____

b _____

c _____

d _____

e _____

f _____

g _____

h _____

2. Matching

Write each of the following terms under the proper heading.

- Right atrium
- Myocardium
- Valve to aorta opens
- Left ventricle
- Valve between left atrium and left ventricle closes
- Pressure in left atrium
- Electrocardiogram
- Endocardium
- Superior vena cava

Anatomy

_____

_____

_____

_____

_____

Physiology

_____

_____

_____

_____

_____

3. Short answer

Briefly describe how the relationship of form and function of a house key and its front door lock are both similar to and different from a chemical messenger and its receptor protein.

_____

_____

_____

4. Section integration

How might a large organism's survival be affected by an inadequate internal circulation network?

_____

_____

1. Short answer

For five different organ systems in the human body, identify a specialized cell type found in that system.

skeletal — bone cells

nervous system — nerve cells or neurons

cardiovascular system — blood cells

muscular system — skeletal muscle cells

reproductive — egg or ovum

2. Concept map

Use each of the following terms once to fill in the blank boxes to correctly complete the extracellular materials and fluids concept map.

- ✓ organs
- ✓ epithelial tissue
- ✓ cells
- ✓ connective tissue
- ✓ muscle tissue
- ✓ nervous tissue
- ✓ organ systems
- ✓ external and internal surfaces
- ✓ matrix
- ✓ glandular secretions
- ✓ bones of the skeleton
- ✓ neuroglia
- ✓ blood
- · materials within digestive tract
- ✓ protein fibers
- ✓ ground substance
- ✓ movement

Extracellular materials and fluids

a. cells

combine to form

Tissues

combine to form

b. organs

interact in

c. organ systems

d. epithelial tissue — covers

e. ext + internal surfaces — and produces

f. glandular secretions

g. connective tissue — contains

Cells

h. matrix — consists of

i. protein fibers

j. ground substance

k. muscle tissue — contracts to produce

l. movement — of

m. bones of skeleton

n. blood

o. materials in digestive tract

p. nervous tissue — consists of

Neurons

q. neuroglia

3. Matching

Order these six levels of organization of the human body from smallest (a) to largest (f).

__3__ tissue   __2__ cell   __4__ organ   __1__ molecule   __6__ organism   __5__ organ system

4. Short answer

Summarize the major functions of each of the following organ systems.

a. skeletal system — support + protection of soft tissue, mineral storage + blood formation

b. digestive system — processing of food + absorption of organic nutrients

c. integumentary system — protection from environmental hazards + temp. control

d. urinary system — elimination of excess water, salts + waste products.

e. nervous system — directing immed. responses to stimuli by coordinating the activities of other organ systems.

3

1. Short answer

For five different organ systems in the human body, identify a specialized cell type found in that system.

_____

_____

_____

_____

_____

2. Concept map

Use each of the following terms once to fill in the blank boxes to correctly complete the extracellular materials and fluids concept map.

- organs
- epithelial tissue
- cells
- connective tissue
- muscle tissue
- nervous tissue
- organ systems
- external and internal surfaces
- matrix
- glandular secretions
- bones of the skeleton
- neuroglia
- blood
- materials within digestive tract
- protein fibers
- ground substance
- movement

Extracellular materials and fluids

a

combine to form

Tissues

combine to form — b — *interact in* — c

d — *covers*

e — *and produces*

f

g — *contains*

Cells

h — *consists of*

i — j

k — *contracts to produce*

l — *of*

m — n — o

p — *consists of*

Neurons

q

3. Matching

Order these six levels of organization of the human body from smallest (a) to largest (f).

_____ tissue   _____ cell   _____ organ   _____ molecule   _____ organism   _____ organ system

4. Short answer

Summarize the major functions of each of the following organ systems.

a skeletal system   _____

b digestive system   _____

c integumentary system   _____

d urinary system   _____

e nervous system   _____

1. Vocabulary

Write the term for each of the following descriptions in the space provided.

a. *pos. feedback* Mechanism that increases a deviation from normal limits after an initial stimulus

b. *neg. feedback* Adjustment of physiological systems to preserve homeostasis

c. *homeostasis* The maintenance of a relatively constant internal environment

d. *pos. feedback* Concentration of hormones circulating in the blood

e. *neg. feedback* Corrective mechanism that opposes or cancels a variation from normal limits

2. Matching

Indicate whether each of the following processes matches with the process of negative feedback or positive feedback.

a. A rise in the level of calcium dissolved in the blood stimulates the release of a hormone that causes bone cells to deposit more of the calcium in bone.

+ feedback

b. Labor contractions become increasingly forceful during childbirth.

– feedback

c. An increase in blood pressure triggers a nervous system response that results in lowering the blood pressure.

– feedback

d. Blood vessel cells damaged by a break in the vessel release chemicals that accelerate the blood clotting process.

+ feedback

3. Short answer

Assuming a normal body temperature range of 36.7°–37.2° C (98°–99° F), identify from the graph below what would happen if there were an increase or decrease in body temperature beyond the normal limits. Use the following descriptive terms to explain what would happen at (a) and (b) on the graph.

- ✓ body surface cools
- ✓ shivering occurs
- ✓ sweating increases
- ✓ temperature declines
- · body heat is conserved
- ✓ blood flow to skin increases
- ✓ blood flow to skin decreases
- ✓ temperature rises

37.8° C /100° F
36.7°–37.2° C /98°–99° F
36.1° C /97° F

a
b

] Normal range

a. *temp. rises*
blood flow to skin increases
sweating increases
body surface cools

b. *temp declines*
blood flow decreases
shivering occurs
body heat is conserved

4. Section integration

It is a warm day and you feel a little chilled. On checking your temperature, you find that your body temperature is 1.5 degrees below normal. Suggest some possible reasons for this situation.

hypothyroid,

5

1. Vocabulary

Write the term for each of the following descriptions in the space provided.

a _____ Mechanism that increases a deviation from normal limits after an initial stimulus

b _____ Adjustment of physiological systems to preserve homeostasis

c _____ The maintenance of a relatively constant internal environment

d _____ Concentration of hormones circulating in the blood

e _____ Corrective mechanism that opposes or cancels a variation from normal limits

2. Matching

Indicate whether each of the following processes matches with the process of negative feedback or positive feedback.

a A rise in the level of calcium dissolved in the blood stimulates the release of a hormone that causes bone cells to deposit more of the calcium in bone.

_____

b Labor contractions become increasingly forceful during childbirth.

_____

c An increase in blood pressure triggers a nervous system response that results in lowering the blood pressure.

_____

d Blood vessel cells damaged by a break in the vessel release chemicals that accelerate the blood clotting process.

_____

3. Short answer

Assuming a normal body temperature range of 36.7°–37.2° C (98°–99° F), identify from the graph below what would happen if there were an increase or decrease in body temperature beyond the normal limits. Use the following descriptive terms to explain what would happen at (a) and (b) on the graph.

- body surface cools
- shivering occurs
- sweating increases
- temperature declines
- body heat is conserved
- blood flow to skin increases
- blood flow to skin decreases
- temperature rises

37.8° C /100° F
36.7°–37.2° C /98°–99° F
36.1° C /97° F

] Normal range

a _____
_____
_____
_____

b _____
_____
_____
_____

4. Section integration

It is a warm day and you feel a little chilled. On checking your temperature, you find that your body temperature is 1.5 degrees below normal. Suggest some possible reasons for this situation.

_____

_____

1. Labeling

Label the directional terms in the figures at right.

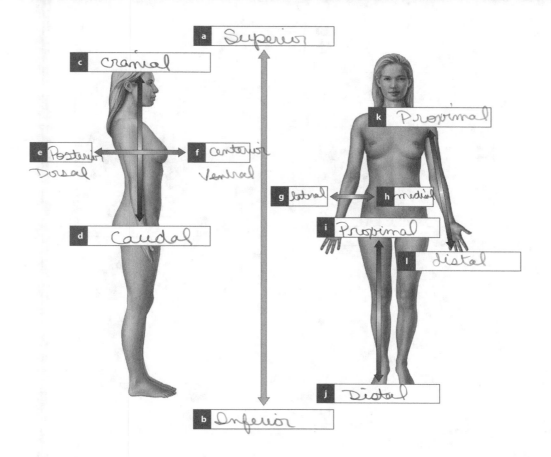

- a Superior
- c Cranial
- e Posterior / Dorsal
- f Anterior / Ventral
- d Caudal
- b Inferior
- k Proximal
- g lateral
- h medial
- i Proximal
- l distal
- j Distal

2. Concept map

Use each of the following terms once to fill in the blank boxes to correctly complete the body cavities concept map.

- digestive glands and organs
- ✓ abdominopelvic cavity
- ✓ thoracic cavity
- ✓ heart
- ✓ mediastinum
- ✓ diaphragm
- ✓ pelvic cavity
- ✓ trachea, esophagus
- ✓ reproductive organs
- ✓ left lung
- peritoneal cavity

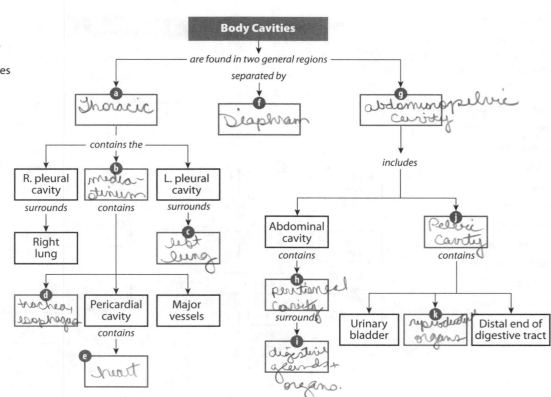

Body Cavities

are found in two general regions
separated by

a Thoracic

f Diaphragm

g abdominopelvic cavity

contains the

R. pleural cavity

b mediastinum

L. pleural cavity

surrounds

contains

surrounds

Right lung

c left lung

includes

Abdominal cavity

j Pelvic Cavity

contains

contains

d trachea, esophagus

Pericardial cavity

Major vessels

h peritoneal cavity

surrounds

Urinary bladder

k reproductive organs

Distal end of digestive tract

contains

e heart

i digestive glands + organs.

1. Labeling

Label the directional terms in the figures at right.

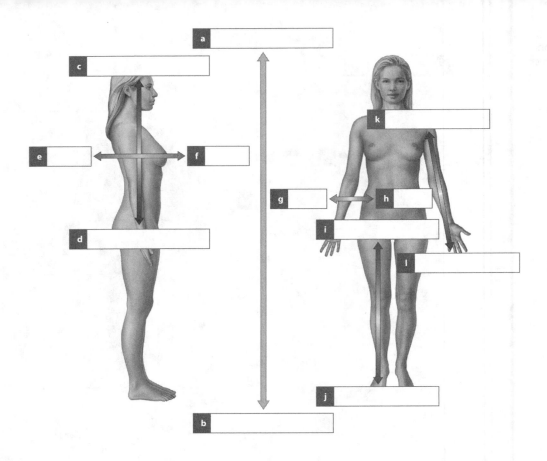

2. Concept map

Use each of the following terms once to fill in the blank boxes to correctly complete the body cavities concept map.

- digestive glands and organs
- abdominopelvic cavity
- thoracic cavity
- heart
- mediastinum
- diaphragm
- pelvic cavity
- trachea, esophagus
- reproductive organs
- left lung
- peritoneal cavity

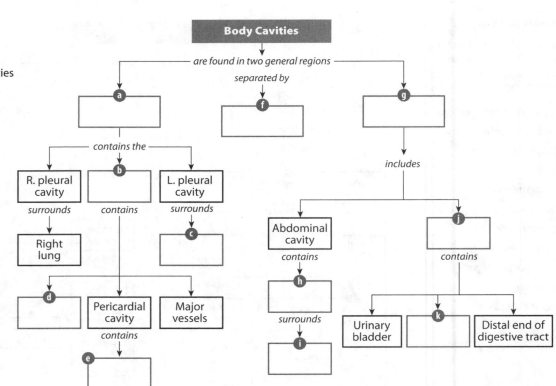

Visual Outline with Key Terms

Summarize the content of each module using the terms in the order provided.

SECTION 1

A&P in Perspective

- homeostasis

1.1

Biology is the study of life

- responsiveness
- adaptability
- growth
- reproduction
- movement
- respiration
- circulation
- digestion
- excretion

1.2

Anatomy is the study of form; physiology is the study of function

- anatomy
- gross anatomy
- macroscopic anatomy
- microscopic anatomy
- physiology

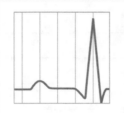

1.3

Form and function are interrelated

- elbow joint
- humerus
- radius
- ulna
- chemical messengers
- receptors

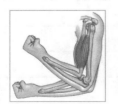

• = *Term boldfaced in this module*

SECTION 2

Levels of Organization

- organism
- organ systems
- organ
- tissue
- cells
- atoms

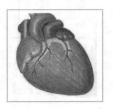

1.4

Cells are the smallest units of life

- cells
- cell theory

1.5

Tissues are specialized groups of cells and cell products

- histology
- primary tissue types
- epithelial tissue
- connective tissue
- muscle tissue
- skeletal muscle tissue
- cardiac muscle tissue
- smooth muscle tissue
- neural tissue
- neurons
- neuroglia
- central nervous system
- peripheral nervous system

• = *Term boldfaced in this module*

1.6

Organs and organ systems perform vital functions

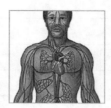

- organ
- organ system
- integumentary system
- skeletal system
- muscular system
- nervous system
- endocrine system
- cardiovascular system
- lymphatic system
- respiratory system
- digestive system
- urinary system
- reproductive system

_____
_____
_____
_____
_____
_____
_____

_____
_____
_____

SECTION 3

Homeostasis

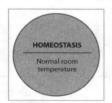

- homeostasis
- homeostatic regulation
- receptor
- control center
- effector
- set point

_____
_____
_____
_____
_____
_____
_____

_____

1.7

Negative feedback provides stability, whereas positive feedback accelerates a process to completion

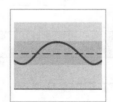

- negative feedback
- positive feedback
- positive feedback loop

_____
_____
_____
_____
_____
_____
_____

● = _Term boldfaced in this module_

SECTION 4

Anatomical Terms

○ eponyms
○ word roots

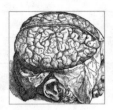

1.8

Superficial anatomy and regional anatomy indicate locations on or in the body

- anatomical position
- supine
- prone
- abdominopelvic quadrants
- abdominopelvic regions

1.9

Directional and sectional terms describe specific points of reference

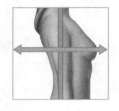

- anterior
- ventral
- posterior
- dorsal
- cranial
- cephalic
- superior
- caudal
- inferior
- medial
- lateral
- proximal
- distal
- superficial
- deep
- transverse section
- horizontal section
- sagittal section
- midsagittal section
- median section
- parasagittal section
- frontal section
- coronal section

● = *Term boldfaced in this module*

1.10

Body cavities protect internal organs and allow them to change shape

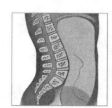

- thoracic cavity
- abdominopelvic cavity
- body cavities
- viscera
- pericardial cavity
- pericardium
- serous membrane
- ventral body cavity
- coelom
- diaphragm
- pleural cavity
- pleura
- mediastinum
- abdominal cavity
- pelvic cavity
- peritoneal cavity
- peritoneum
- retroperitoneal

• = *Term boldfaced in this module*

1. Short answer

Fill in the missing information in the following table.

Element	Number of protons	Number of electrons	Number of neutrons	Mass number
Helium	a	2	2	b
Hydrogen	1	c	d	1
Carbon	6	e	6	f
Nitrogen	g	7	h	14
Calcium	i	j	20	40

2. Short answer

Indicate which of the following molecules are compounds and which are elements.

H_2 (hydrogen)　　　H_2O (water)　　　O_2 (oxygen)　　　CO (carbon monoxide)

a _____　b _____　c _____　d _____

3. Matching

Match the following terms with the most closely related description.

- atomic number
- electrons
- protons
- neutrons
- isotopes
- ions
- ionic bond
- covalent bond
- mass number
- element
- compound
- hydrogen bond

a _____ Atoms that have gained or lost electrons

b _____ Located in the nucleus, have no charge

c _____ Atoms of two or more different elements bonded together

d _____ The number of protons in an atom

e _____ Attractive force between water molecules

f _____ Type of chemical bond within a water molecule

g _____ The number of subatomic particles in the nucleus

h _____ Substance composed only of atoms with same atomic number

i _____ Subatomic particles in the nucleus, have charge

j _____ Atoms of the same element with different masses

k _____ Type of chemical bond in table salt

l _____ Subatomic particles outside the nucleus, have charge

4. Section integration

Describe how the following pairs of terms concerning atomic interactions are similar and how they are different.

a inert element/reactive element _____

b polar molecules/nonpolar molecules _____

c covalent bond/ionic bond _____

1. Short answer

Fill in the missing information in the following table.

Element	Number of protons	Number of electrons	Number of neutrons	Mass number
Helium	a	2	2	b
Hydrogen	1	c	d	1
Carbon	6	e	6	f
Nitrogen	g	7	h	14
Calcium	i	j	20	40

2. Short answer

Indicate which of the following molecules are compounds and which are elements.

H_2 (hydrogen)　　　H_2O (water)　　　O_2 (oxygen)　　　CO (carbon monoxide)

a _____　　b _____　　c _____　　d _____

3. Matching

Match the following terms with the most closely related description.

- atomic number
- electrons
- protons
- neutrons
- isotopes
- ions
- ionic bond
- covalent bond
- mass number
- element
- compound
- hydrogen bond

a _____ Atoms that have gained or lost electrons

b _____ Located in the nucleus, have no charge

c _____ Atoms of two or more different elements bonded together

d _____ The number of protons in an atom

e _____ Attractive force between water molecules

f _____ Type of chemical bond within a water molecule

g _____ The number of subatomic particles in the nucleus

h _____ Substance composed only of atoms with same atomic number

i _____ Subatomic particles in the nucleus, have charge

j _____ Atoms of the same element with different masses

k _____ Type of chemical bond in table salt

l _____ Subatomic particles outside the nucleus, have charge

4. Section integration

Describe how the following pairs of terms concerning atomic interactions are similar and how they are different.

a inert element/reactive element _____

b polar molecules/nonpolar molecules _____

c covalent bond/ionic bond _____

1. Short answer

Using chemical notation, write the formula of each of the following.

a One molecule of hydrogen _____

b Two atoms of hydrogen _____

c Six molecules of water _____

d One molecule of sucrose (in this order: 12 atoms of carbon, 22 atoms of hydrogen, and 11 atoms of oxygen) _____

2. Short answer

Write the chemical equation for the following chemical reaction: one molecule of glucose combined with six molecules of oxygen produce six molecules of carbon dioxide and six molecules of water.

_____

3. Short answer

Indicate which of the following reactions is a hydrolysis reaction, and which is a dehydration synthesis reaction.

a   _____

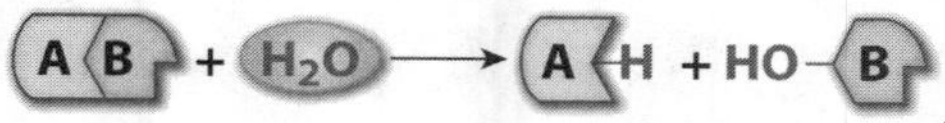

b   _____

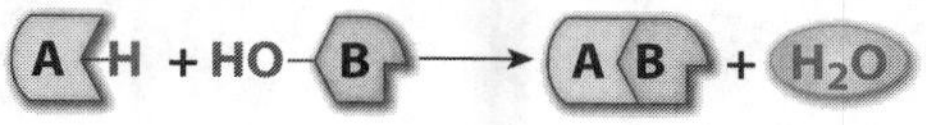

4. Matching

Match the following terms with the most closely related description.

- exergonic
- activation energy
- organic compounds
- exchange reaction
- hydrolysis
- endergonic
- reactants
- enzyme

a _____ Catalyst

b _____ Starting substances in a chemical reaction

c _____ Chemical reaction involving water

d _____ Reactions that absorb energy

e _____ Shuffles parts of reactants

f _____ Primary components are carbon and hydrogen

g _____ Reactions that release energy

h _____ Requirement for starting a chemical reaction

5. Section integration

In a metabolic pathway that consists of four steps, how would decreasing the amount of enzyme that catalyzes the second step affect the amount of product produced at the end of the pathway?

_____

_____

1. Short answer

Using chemical notation, write the formula of each of the following.

a One molecule of hydrogen _____

b Two atoms of hydrogen _____

c Six molecules of water _____

d One molecule of sucrose (in this order: 12 atoms of carbon, 22 atoms of hydrogen,

and 11 atoms of oxygen) _____

2. Short answer

Write the chemical equation for the following chemical reaction: one molecule of glucose combined with six molecules of oxygen produce six molecules of carbon dioxide and six molecules of water.

_____

3. Short answer

Indicate which of the following reactions is a hydrolysis reaction, and which is a dehydration synthesis reaction.

a $A\text{-}B + H_2O \longrightarrow A\text{-}H + HO\text{-}B$ _____

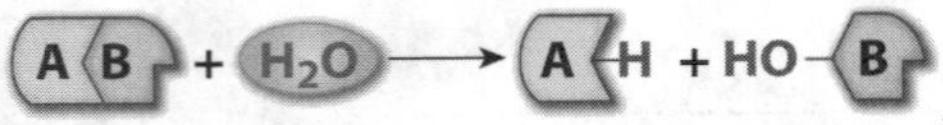

b $A\text{-}H + HO\text{-}B \longrightarrow A\text{-}B + H_2O$ _____

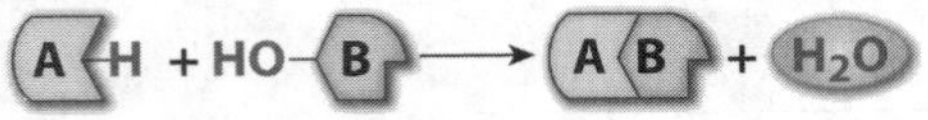

4. Matching

Match the following terms with the most closely related description.

- exergonic
- activation energy
- organic compounds
- exchange reaction
- hydrolysis
- endergonic
- reactants
- enzyme

a _____ Catalyst

b _____ Starting substances in a chemical reaction

c _____ Chemical reaction involving water

d _____ Reactions that absorb energy

e _____ Shuffles parts of reactants

f _____ Primary components are carbon and hydrogen

g _____ Reactions that release energy

h _____ Requirement for starting a chemical reaction

5. Section integration

In a metabolic pathway that consists of four steps, how would decreasing the amount of enzyme that catalyzes the second step affect the amount of product produced at the end of the pathway?

_____

_____

1. Short answer

List four properties of water important to the functioning of the human body.

a _____

b _____

c _____

d _____

2. Matching

Match the following terms with the most closely related description.

- solvent
- water
- buffers
- hydrophilic
- inorganic compounds
- hydrophobic
- acid
- solute
- alkaline
- salt

a _____ HCl, NaOH, and NaCl

b _____ A dissolved substance

c _____ A solution with a pH greater than 7

d _____ Molecules that readily interact with water

e _____ Fluid medium of a solution

f _____ Ionic compound not containing hydrogen ions or hydroxide ions

g _____ Compounds that stabilize pH in body fluids

h _____ Solution with a pH of 6.5

i _____ Molecules that do not interact with water

j _____ Makes up two-thirds of human body weight

3. Short answer

Identify the regions a–c on the pH scale below.

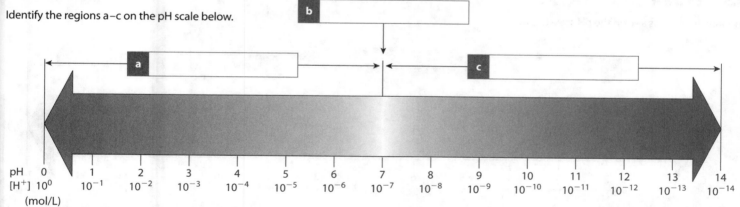

pH	0	1	2	3	4	5	6	7	8	9	10	11	12	13	14
$[H^+]$ (mol/L)	10^0	10^{-1}	10^{-2}	10^{-3}	10^{-4}	10^{-5}	10^{-6}	10^{-7}	10^{-8}	10^{-9}	10^{-10}	10^{-11}	10^{-12}	10^{-13}	10^{-14}

d How much more or less acidic is a solution of pH 3 compared to one with a pH of 6? _____

e Describe three negative effects of abnormal pH fluctuations in the human body.

_____

4. Section integration

The addition of table salt to pure water does not result in a change in its pH. Why?

_____

_____

1. Short answer

List four properties of water important to the functioning of the human body.

a _____

b _____

c _____

d _____

2. Matching

Match the following terms with the most closely related description.

- solvent
- water
- buffers
- hydrophilic
- inorganic compounds
- hydrophobic
- acid
- solute
- alkaline
- salt

a _____ HCl, NaOH, and NaCl

b _____ A dissolved substance

c _____ A solution with a pH greater than 7

d _____ Molecules that readily interact with water

e _____ Fluid medium of a solution

f _____ Ionic compound not containing hydrogen ions or hydroxide ions

g _____ Compounds that stabilize pH in body fluids

h _____ Solution with a pH of 6.5

i _____ Molecules that do not interact with water

j _____ Makes up two-thirds of human body weight

3. Short answer

Identify the regions a–c on the pH scale below.

b

a

c

pH	0	1	2	3	4	5	6	7	8	9	10	11	12	13	14
$[H^+]$ (mol/L)	10^0	10^{-1}	10^{-2}	10^{-3}	10^{-4}	10^{-5}	10^{-6}	10^{-7}	10^{-8}	10^{-9}	10^{-10}	10^{-11}	10^{-12}	10^{-13}	10^{-14}

d How much more or less acidic is a solution of pH 3 compared to one with a pH of 6? _____

e Describe three negative effects of abnormal pH fluctuations in the human body.

_____

4. Section integration

The addition of table salt to pure water does not result in a change in its pH. Why?

_____

_____

1. Matching

Match the following terms with the most closely related description.

- monosaccharide
- ATP
- polyunsaturated
- glycerol
- cholesterol
- isomers
- glycogen
- active site
- nucleotide
- RNA
- peptide

a _____ Polysaccharide with an energy-storage role in animal tissues

b _____ Molecules with same chemical formula but different structure

c _____ A fatty acid with more than one C-to-C double covalent bond

d _____ The region of an enzyme that binds the substrate

e _____ Three-carbon molecule that combines with fatty acids

f _____ A steroid essential to plasma membranes

g _____ A high-energy compound consisting of adenosine and three phosphate groups

h _____ A nucleic acid that contains the sugar ribose

i _____ The covalent bond between the carboxylic acid and amino groups of adjacent amino acids

j _____ Organic molecule consisting of a sugar, a phosphate group, and a nitrogenous base

k _____ A simple sugar

2. Vocabulary

In the space provided, write the boldfaced terms introduced in this section that contain the indicated word part.

Word Part	Meaning	Terms
a poly-	many	_____
b tri-	three	_____
c di-	two	_____
d glyco-	sugar	_____

3. Concept map

Use each of the following terms once to fill in the blank boxes to correctly complete the organic compounds concept map.

- lipids
- carbohydrates
- nucleic acids
- disaccharides
- RNA
- fatty acids
- phosphate groups
- glycerol
- polysaccharides
- proteins
- monosaccharides
- ATP
- amino acids
- DNA
- nucleotides

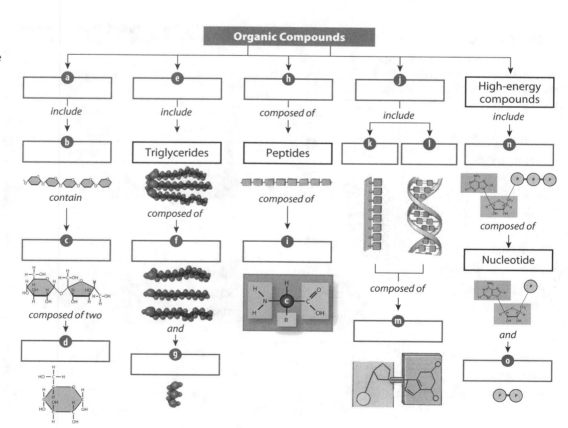

21

1. Matching

Match the following terms with the most closely related description.

- monosaccharide
- ATP
- polyunsaturated
- glycerol
- cholesterol
- isomers
- glycogen
- active site
- nucleotide
- RNA
- peptide

a _____ Polysaccharide with an energy-storage role in animal tissues

b _____ Molecules with same chemical formula but different structure

c _____ A fatty acid with more than one C-to-C double covalent bond

d _____ The region of an enzyme that binds the substrate

e _____ Three-carbon molecule that combines with fatty acids

f _____ A steroid essential to plasma membranes

g _____ A high-energy compound consisting of adenosine and three phosphate groups

h _____ A nucleic acid that contains the sugar ribose

i _____ The covalent bond between the carboxylic acid and amino groups of adjacent amino acids

j _____ Organic molecule consisting of a sugar, a phosphate group, and a nitrogenous base

k _____ A simple sugar

2. Vocabulary

In the space provided, write the boldfaced terms introduced in this section that contain the indicated word part.

Word Part		Meaning	Terms
a	poly-	many	_____
b	tri-	three	_____
c	di-	two	_____
d	glyco-	sugar	_____

3. Concept map

Use each of the following terms once to fill in the blank boxes to correctly complete the organic compounds concept map.

- lipids
- carbohydrates
- nucleic acids
- disaccharides
- RNA
- fatty acids
- phosphate groups
- glycerol
- polysaccharides
- proteins
- monosaccharides
- ATP
- amino acids
- DNA
- nucleotides

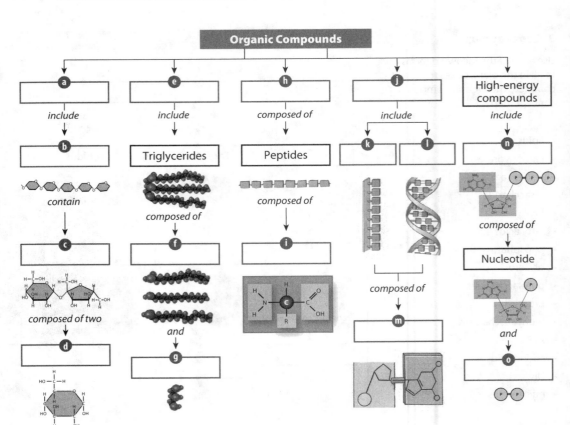

Visual Outline with Key Terms

Summarize the content of each module using the terms in the order provided.

SECTION 1

Atoms and Molecules

- mass
- atoms
- subatomic particles
- protons
- neutrons
- electrons
- nucleus
- electron cloud
- molecule

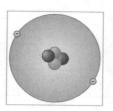

_____

_____

_____

_____

_____

_____

2.1

Typical atoms contain protons, neutrons, and electrons

- atomic number
- mass number
- element
- electron cloud
- electron shell
- isotopes
- atomic weight
- dalton
- trace elements
- chemical symbol

_____

_____

_____

_____

_____

_____

2.2

Electrons occupy various energy levels

- inert
- inert gases
- reactive
- cation
- anion
- chemical bonds

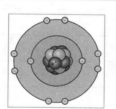

_____

_____

_____

_____

_____

• = _Term boldfaced in this module_

2.3

The most common chemical bonds are ionic bonds and covalent bonds

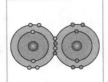

- compound
- ionic bonds
- covalent bonds
- molecule
- single covalent bond

- double covalent bond
- nonpolar molecules
- polar molecule
- polar covalent bonds

2.4

Matter may exist as a solid, a liquid, or a gas

- hydrogen bond
- surface tension

SECTION 2

Chemical Reactions

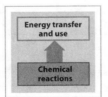

- reactants
- products
- metabolism
- work

- energy
- kinetic energy
- potential energy

● = Term boldfaced in this module

2.5

Chemical notation is a concise method of describing chemical reactions

- chemical notation
- balanced
- mole (mol)
- molecular weight

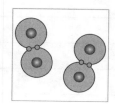

_____

_____

_____

_____

_____

_____

_____

_____

2.6

There are three basic types of chemical reactions

- decomposition
- hydrolysis
- catabolism
- synthesis
- dehydration synthesis
- anabolism
- equilibrium
- exchange reaction

_____

_____

_____

_____

_____

_____

_____

_____

2.7

Enzymes lower the activation energy requirements of chemical reactions

- activation energy
- enzymes
- catalysts
- metabolic pathway
- exergonic
- endergonic
- metabolites
- nutrients
- organic compounds
- inorganic compounds

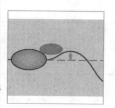

_____

_____

_____

_____

_____

_____

_____

_____

_____

• = _Term boldfaced in this module_

SECTION 3

The Importance of Water in the Body

- lubrication
- reactivity
- heat capacity
- thermal inertia
- solubility
- solution
- solvent
- solutes
- aqueous solutions

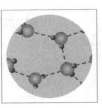

_____

_____

_____

_____

_____

_____

_____

2.8

Physiological systems depend on water

- ionization
- dissociation
- hydration sphere
- hydrophilic
- electrolytes
- colloid
- suspension
- hydrophobic

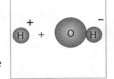

_____

_____

_____

_____

_____

_____

_____

2.9

Regulation of body fluid pH is vital for homeostasis

- hydrogen ion (H^+)
- hydroxide ion (OH^-)
- pH
- acidosis
- alkalosis
- acidic
- neutral
- alkaline
- acid
- hydrochloric acid
- base
- sodium hydroxide
- carbonic acid
- bicarbonate ion
- salt
- buffers
- buffer systems

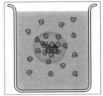

_____

_____

_____

_____

_____

_____

_____

_____

_____

• = *Term boldfaced in this module*

Organic Compounds

- organic compounds
- functional groups
- carboxylic acid group
- amino group
- hydroxyl group
- phosphate group
- high-energy compounds

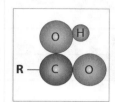

2.10

Carbohydrates contain carbon, hydrogen, and oxygen, usually in a 1:2:1 ratio

- carbohydrate
- monosaccharide
- glucose
- isomers
- fructose
- disaccharide
- sucrose
- polysaccharides
- starches
- glycogen

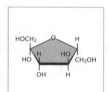

2.11

Lipids often contain a carbon-to-hydrogen ratio of 1:2

- lipids
- fatty acids
- carboxylic acid group
- saturated fatty acid
- unsaturated fatty acid
- glycerol
- glyceride
- monoglyceride
- diglyceride
- triglyceride

● = *Term boldfaced in this module*

2.12

Eicosanoids, steroids, phospholipids, and glycolipids have diverse functions

- structural lipids
- eicosanoids
- leukotrienes
- prostaglandins
- steroids
- cholesterol
- phospholipids
- glycolipids
- micelles

_____

_____

2.13

Proteins are formed from amino acids

- proteins
- amino acids
- peptide bond
- peptides
- dipeptide
- polypeptides
- primary structure
- secondary structure
- tertiary structure
- quaternary structure
- hemoglobin
- keratin
- collagen
- denaturation

_____

_____

2.14

Enzymes are proteins with important regulatory functions

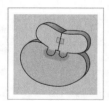

- substrates
- active site
- enzyme-substrate complex
- saturation limit

_____

_____

• = _Term boldfaced in this module_

High-energy compounds may store and transfer a portion of energy released during enzymatic reactions

- high-energy compound
- high-energy bonds
- adenosine triphosphate (ATP)
- adenosine monophosphate (AMP)
- adenosine diphosphate (ADP)

DNA and RNA are nucleic acids

- nucleic acids
- deoxyribonucleic acid (DNA)
- ribonucleic acid (RNA)
- nucleotides
- nitrogenous base
- purine
- pyrimidine
- adenine
- guanine
- cytosine
- thymine
- uracil
- complementary base pairs
- complementary strands
- messenger RNA (mRNA)
- transfer RNA (tRNA)
- ribosomal RNA (rRNA)

• = Term boldfaced in this module

1. Short answer

Correctly label the indicated structures on the accompanying diagram of a cell and then describe the functions of each.

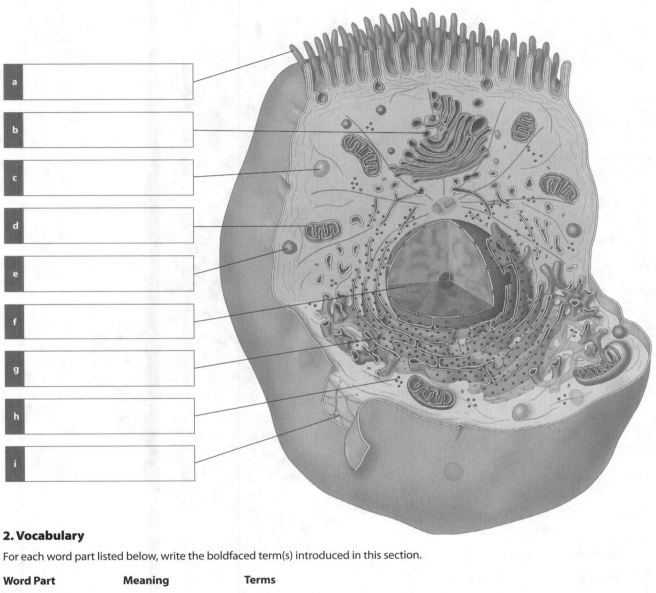

a

b

c

d

e

f

g

h

i

2. Vocabulary

For each word part listed below, write the boldfaced term(s) introduced in this section.

Word Part	Meaning	Terms
a glycos-	sugar	_____
b aero-	air	_____
c micro-	small	_____
d lyso-	a loosening	_____

3. Section integration

What is the advantage of having some of the cellular organelles enclosed by a membrane similar to the plasma membrane? _____

_____

1. Short answer

Correctly label the indicated structures on the accompanying diagram of a cell and then describe the functions of each.

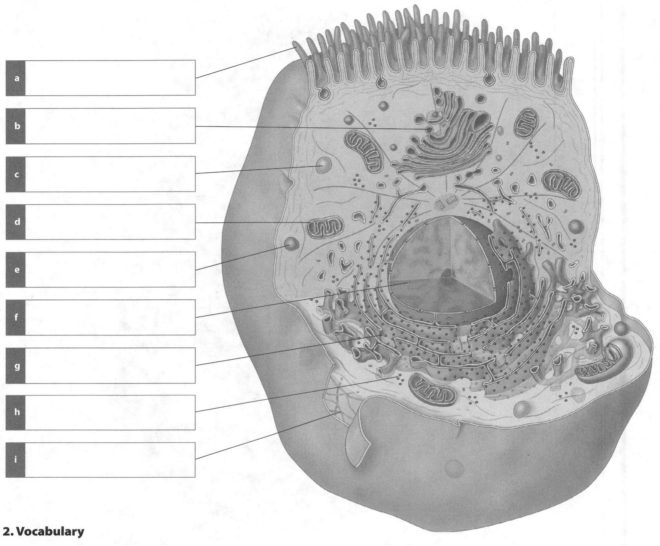

a	
b	
c	
d	
e	
f	
g	
h	
i	

2. Vocabulary

For each word part listed below, write the boldfaced term(s) introduced in this section.

Word Part	Meaning	Terms
a glycos-	sugar	
b aero-	air	
c micro-	small	
d lyso-	a loosening	

3. Section integration

What is the advantage of having some of the cellular organelles enclosed by a membrane similar to the plasma membrane?

1. Matching

Match the following terms with the most closely related description.

- introns
- transcription
- tRNA
- chromosomes
- exons
- genetic information
- nucleus
- thymine
- mRNA
- gene
- uracil
- nuclear envelope
- nuclear pore
- nucleoli

a	_____	DNA strands and histones
b	_____	DNA nitrogen base
c	_____	double membrane
d	_____	mRNA noncoding regions
e	_____	RNA nitrogen base
f	_____	assemble ribosomal subunits
g	_____	passageway for functional mRNA
h	_____	mRNA formation
i	_____	functional unit of heredity
j	_____	mRNA coding regions
k	_____	codon
l	_____	anticodon
m	_____	cell control center
n	_____	DNA nucleotide sequence

2. Short answer

This is a sequence of DNA bases in a protein-coding gene.

T A C A A A A C A C G G C G G A A T

a Provide the corresponding mRNA base sequence and insert a slash mark (/) between the codons:

_____

b Convert the mRNA codons above to tRNA anticodons:

_____

c Using the triplet code table in Module 3.8, translate the anticodon sequence into the amino acid sequence of this polypeptide:

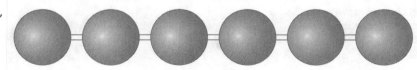

3. Section integration

The nucleus is often described as the control center of the cell. Explain the role of the nucleus in maintaining homeostasis.

_____
_____
_____

1. Matching

Match the following terms with the most closely related description.

- introns
- transcription
- tRNA
- chromosomes
- exons
- genetic information
- nucleus
- thymine
- mRNA
- gene
- uracil
- nuclear envelope
- nuclear pore
- nucleoli

a _____ DNA strands and histones

b _____ DNA nitrogen base

c _____ double membrane

d _____ mRNA noncoding regions

e _____ RNA nitrogen base

f _____ assemble ribosomal subunits

g _____ passageway for functional mRNA

h _____ mRNA formation

i _____ functional unit of heredity

j _____ mRNA coding regions

k _____ codon

l _____ anticodon

m _____ cell control center

n _____ DNA nucleotide sequence

2. Short answer

This is a sequence of DNA bases in a protein-coding gene.

T A C A A A A C A C G G C G G A A T

a Provide the corresponding mRNA base sequence and insert a slash mark (/) between the codons:

_____

b Convert the mRNA codons above to tRNA anticodons:

_____

c Using the triplet code table in Module 3.8, translate the anticodon sequence into the amino acid sequence of this polypeptide:

3. Section integration

The nucleus is often described as the control center of the cell. Explain the role of the nucleus in maintaining homeostasis.

_____

_____

_____

1. Concept map

Use each of the following terms once to fill in the blank boxes to correctly complete the membrane permeability concept map.

- exocytosis
- diffusion
- "cell eating"
- molecular size
- pinocytosis
- facilitated diffusion
- vesicular transport
- net diffusion of water
- active transport
- specificity

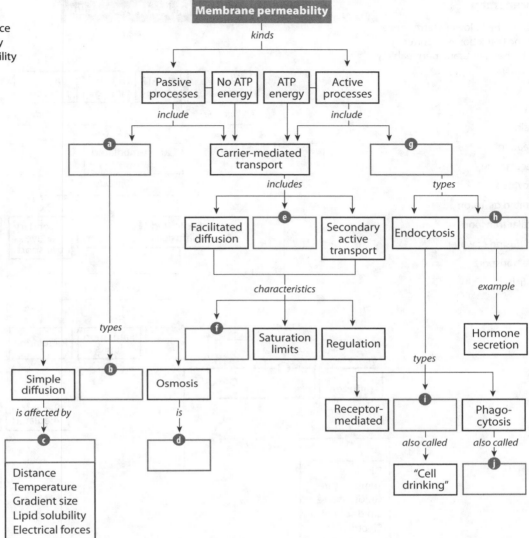

2. Short answer

Classify each of the following situations as an example of diffusion, osmosis, or neither.

- You walk into a room and smell a balsam-scented candle.

- A drop of food coloring disperses within a liquid medium.

- Water flows through a garden hose.

- A sugar cube placed in a cup of hot tea dissolves.

- Grass in the yard wilts after being exposed to excess chemical fertilizer.

- After soaking several hours in water containing sodium chloride, a stalk of celery weighs less than before it was placed in the salty water.

a   _____

b   _____

c   _____

d   _____

e   _____

f   _____

1. Concept map

Use each of the following terms once to fill in the blank boxes to correctly complete the membrane permeability concept map.

- exocytosis
- diffusion
- "cell eating"
- molecular size
- pinocytosis
- facilitated diffusion
- vesicular transport
- net diffusion of water
- active transport
- specificity

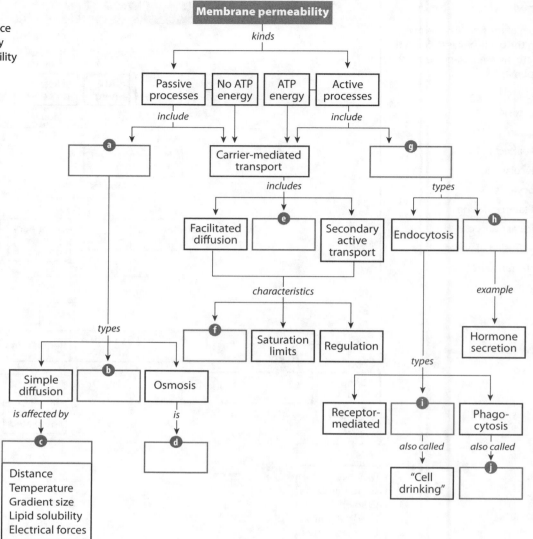

Membrane permeability

kinds

Passive processes | No ATP energy | ATP energy | Active processes

include ... *include*

a ... Carrier-mediated transport ... g

includes ... *types*

Facilitated diffusion ... e ... Secondary active transport ... Endocytosis ... h

characteristics

f ... Saturation limits ... Regulation

types

Simple diffusion ... b ... Osmosis

is affected by ... *is*

c ... d

Distance
Temperature
Gradient size
Lipid solubility
Electrical forces

example

Hormone secretion

types

Receptor-mediated ... i ... Phago-cytosis

also called ... *also called*

"Cell drinking" ... j

2. Short answer

Classify each of the following situations as an example of diffusion, osmosis, or neither.

- You walk into a room and smell a balsam-scented candle.

- A drop of food coloring disperses within a liquid medium.

- Water flows through a garden hose.

- A sugar cube placed in a cup of hot tea dissolves.

- Grass in the yard wilts after being exposed to excess chemical fertilizer.

- After soaking several hours in water containing sodium chloride, a stalk of celery weighs less than before it was placed in the salty water.

a _____

b _____

c _____

d _____

e _____

f _____

1. Concept map

Use each of the following terms once to fill in the blank boxes to correctly complete the cell cycle concept map.

- metaphase
- DNA replication
- somatic cells
- G_2 phase
- telophase
- mitosis
- cytokinesis
- G_1 phase

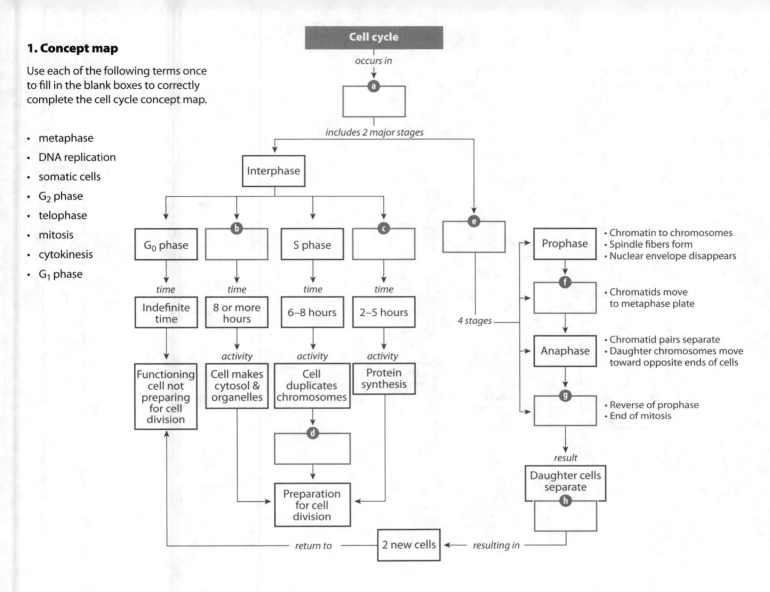

2. Vocabulary

In the space provided, write the boldfaced terms introduced in this section that contain the indicated word part.

Word Part		Meaning	Terms
a	telo-	end	_____
b	pro-	before	_____
c	centro-	in the middle	_____

3. Section integration

The muscle cells that that make up skeletal muscle tissue are large and multinucleated. They form early in development as groups of embryonic cells fuse together, contributing their nuclei and losing their individual plasma membranes. Describe an alternate mechanism that would also result in a large, multinucleated cell. _____

_____

_____

1. Concept map

Use each of the following terms once to fill in the blank boxes to correctly complete the cell cycle concept map.

- metaphase
- DNA replication
- somatic cells
- G_2 phase
- telophase
- mitosis
- cytokinesis
- G_1 phase

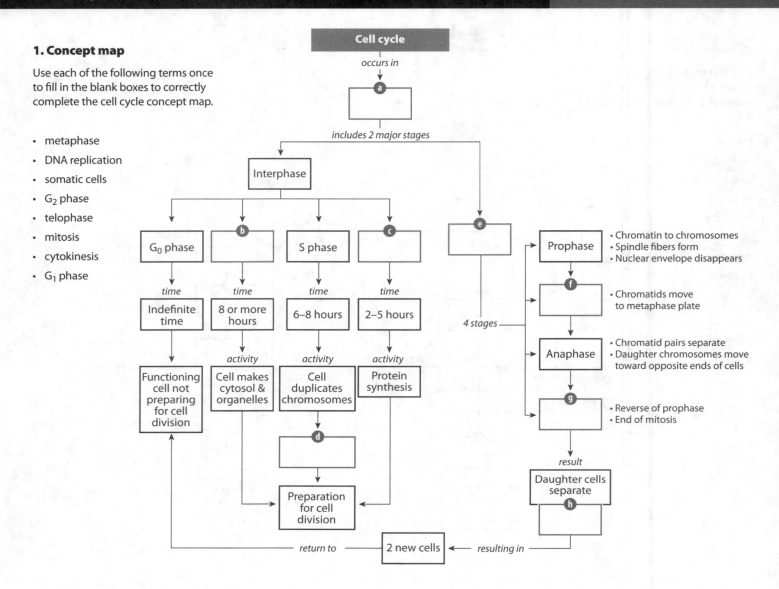

2. Vocabulary

In the space provided, write the boldfaced terms introduced in this section that contain the indicated word part.

Word Part		Meaning	Terms
a	telo-	end	_____
b	pro-	before	_____
c	centro-	in the middle	_____

3. Section integration

The muscle cells that that make up skeletal muscle tissue are large and multinucleated. They form early in development as groups of embryonic cells fuse together, contributing their nuclei and losing their individual plasma membranes. Describe an alternate mechanism that would also result in a large, multinucleated cell. _____

_____

Visual Outline with Key Terms

Summarize the content of each module using the terms in the order provided.

Atoms and Molecules

- ○ cell theory
- • differentiation

_____

_____

_____

_____

_____

_____

_____

3.1

Cells are the smallest living units of life

- extracellular fluid
- interstitial fluid
- plasma membrane
- cytoplasm
- cytosol
- organelles
- nonmembranous organelles
- membranous organelles
- peroxisome
- lysosome
- microvilli
- Golgi apparatus
- nucleus
- nuclear envelope
- endoplasmic reticulum (ER)
- smooth ER
- rough ER
- ribosomes
- cytoskeleton
- centrosome
- mitochondria

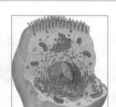

_____

_____

_____

_____

_____

_____

_____

_____

_____

3.2

The plasma membrane isolates the cell from its environment and performs varied functions

- plasma membrane
- glycocalyx
- integral proteins
- transmembrane proteins
- peripheral proteins
- phospholipid bilayer
- anchoring proteins
- recognition proteins
- receptor proteins
- ligands
- carrier proteins
- channels

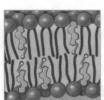

_____

_____

_____

_____

_____

_____

_____

• = _Term boldfaced in this module_

3.3

The cytoskeleton plays both a structural and a functional role

- cytoskeleton
- microvilli
- microfilaments
- actin
- terminal web
- intermediate filaments
- microtubules
- thick filaments
- centrioles
- centrosome
- cilia
- basal body

3.4

Ribosomes are responsible for protein synthesis and are often associated with the endoplasmic reticulum

- ribosomes
- ribosomal RNA (rRNA)
- endoplasmic reticulum (ER)
- cisternae
- smooth endoplasmic reticulum (SER)
- rough endoplasmic reticulum (RER)
- fixed ribosomes
- transport vesicles

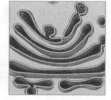

3.5

The Golgi apparatus is a packaging center

- Golgi apparatus
- cisternae
- transport vesicles
- forming face
- maturing face
- membrane renewal vesicles
- secretory vesicles
- lysosomes
- autolysis
- membrane flow

• = *Term boldfaced in this module*

3.6

Mitochondria are the powerhouses of the cell

- mitochondria
- matrix
- cristae
- glycolysis
- citric acid cycle
- aerobic metabolism

SECTION 2

Structure and Function of the Nucleus

- nucleus
- cellular-level homeostasis
- short-term adjustments
- long-term adjustments

3.7

The nucleus contains DNA, RNA, organizing proteins, and enzymes

- nucleus
- nuclear envelope
- perinuclear space
- nuclear pores
- nucleoplasm
- nucleoli
- histones
- nucleosomes
- chromatin
- chromosomes
- centromere

3.8

Protein synthesis involves DNA, enzymes, and three types of RNA

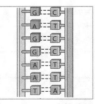

- genetic code
- gene
- triplet code
- triplet
- gene activation
- messenger RNA (mRNA)
- codons
- anticodons
- transfer RNA (tRNA)
- ribosomal RNA (rRNA)

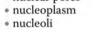 = Term boldfaced in this module

3.9

Transcription encodes genetic instructions on a strand of RNA

- transcription
- control segment
- template strand
- RNA polymerase
- immature mRNA
- introns
- exons

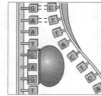

3.10

Translation builds polypeptides as directed by an mRNA strand

- translation
- anticodon

SECTION 3

How Things Enter and Leave the Cell

- permeability
- freely permeable membranes
- selectively permeable membranes
- impermeable membranes

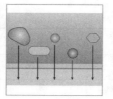

3.11

Diffusion is movement driven by concentration differences

- diffusion
- gradient

• = *Term boldfaced in this module*

3.12

Osmosis is the passive movement of water

- osmosis
- osmotic flow
- osmotic pressure
- hydrostatic pressure
- osmolarity (osmotic concentration)
- tonicity
- isotonic
- hypotonic
- hemolysis
- hypertonic
- crenation
- normal saline

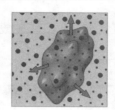

3.13

In carrier-mediated transport, integral proteins facilitate membrane passage

- carrier-mediated transport
- carrier proteins
- cotransport
- countertransport
- exchange pump
- facilitated diffusion
- active transport
- ion pumps
- sodium–potassium ATPase
- secondary active transport

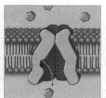

3.14

In vesicular transport, vesicles perform selective membrane passage

- vesicular transport
- endocytosis
- exocytosis
- receptor-mediated endocytosis
- endosomes
- ligand
- coated vesicles
- pinocytosis
- phagocytosis
- phagosomes
- phagocytes
- macrophages
- pseudopodia

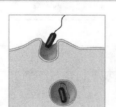

• = *Term boldfaced in this module*

SECTION 4

The Cell Life Cycle

- cell division
- apoptosis
- mitosis
- meiosis
- daughter cells
- interphase
- cytokinesis

3.15

During interphase, the cell prepares for cell division

- somatic cells
- interphase
- G$_1$ phase
- S phase
- G$_2$ phase
- G$_0$ phase
- stem cells
- DNA replication
- DNA polymerase
- ligases

3.16

Mitosis distributes chromosomes before cytokinesis separates the daughter cells

- mitosis
- cytokinesis
- prophase
- chromatid
- kinetochore
- astral rays
- spindle fibers
- metaphase
- metaphase plate
- anaphase
- spindle apparatus
- telophase
- cleavage furrow

3.17

Tumors and cancer are characterized by abnormal cell growth and division

- cancer
- mutations
- tumor
- neoplasm
- benign tumor
- malignant tumor
- metastasis
- invasion

• = *Term boldfaced in this module*

1. Labeling

Label the types of epithelial tissues shown in the drawing to the right.

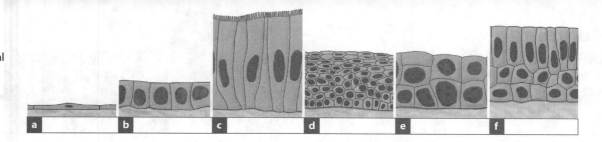

2. Short answer

Complete the following table by writing the missing epithelium type or structure.

Type of Epithelium	Structure (or Organ)
a	Lining of the trachea
Transitional epithelium	b
c	Surface of the skin
d	Lining of the small intestine
Simple squamous epithelium	e
Simple cuboidal epithelium	f
g	Ducts of sweat glands and mammary glands

3. Concept map

Use the following terms to fill in the blank spaces to complete the glandular epithelia concept map.

- endocrine glands
- ducts
- mucous cells
- apocrine secretion
- interstitial fluid
- mucus
- merocrine secretion
- epithelial surfaces
- mucin
- exocrine glands
- holocrine secretion

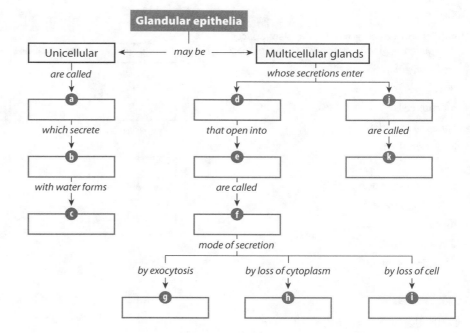

4. Vocabulary

Fill in the blanks with the appropriate term.

a _____ A term meaning no blood vessels

b _____ A gland whose glandular cells form blind pockets

c _____ A type of epithelium that withstands stretching and that changes in appearance as stretching occurs

d _____ The cell junction formed by the partial fusion of the lipid portions of two plasma membranes

e _____ The complex structure attached to the basal surface of an epithelium

f _____ A gland that has a single duct

g _____ The type of epithelium lining the subdivisions of the ventral body cavity

1. Labeling

Label the types of epithelial tissues shown in the drawing to the right.

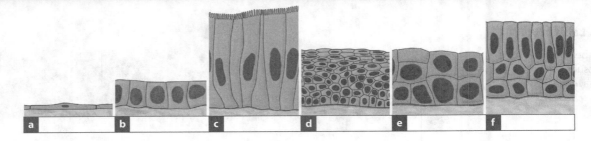

| a | b | c | d | e | f |

2. Short answer

Complete the following table by writing the missing epithelium type or structure.

Type of Epithelium	Structure (or Organ)
a	Lining of the trachea
Transitional epithelium	b
c	Surface of the skin
d	Lining of the small intestine
Simple squamous epithelium	e
Simple cuboidal epithelium	f
g	Ducts of sweat glands and mammary glands

3. Concept map

Use the following terms to fill in the blank spaces to complete the glandular epithelia concept map.

- endocrine glands
- ducts
- mucous cells
- apocrine secretion
- interstitial fluid
- mucus
- merocrine secretion
- epithelial surfaces
- mucin
- exocrine glands
- holocrine secretion

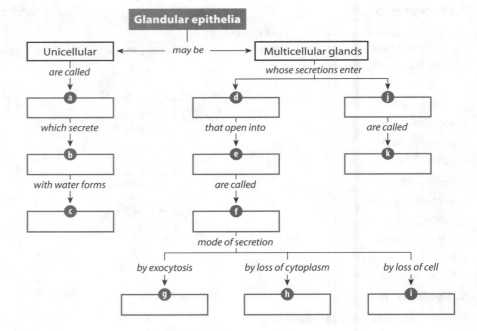

Glandular epithelia

Unicellular ← may be → Multicellular glands

are called

a

which secrete

b

with water forms

c

whose secretions enter

d

that open into

e

are called

f

mode of secretion

j

are called

k

by exocytosis

g

by loss of cytoplasm

h

by loss of cell

i

4. Vocabulary

Fill in the blanks with the appropriate term.

a _____ A term meaning no blood vessels

b _____ A gland whose glandular cells form blind pockets

c _____ A type of epithelium that withstands stretching and that changes in appearance as stretching occurs

d _____ The cell junction formed by the partial fusion of the lipid portions of two plasma membranes

e _____ The complex structure attached to the basal surface of an epithelium

f _____ A gland that has a single duct

g _____ The type of epithelium lining the subdivisions of the ventral body cavity

1. Concept map

Use the following terms once to fill in the blank boxes to complete the connective tissue concept map.

- loose connective tissues
- chondrocytes in lacunae
- fluid connective tissue
- tendons
- ligaments
- regular
- hyaline
- blood
- adipose
- bone

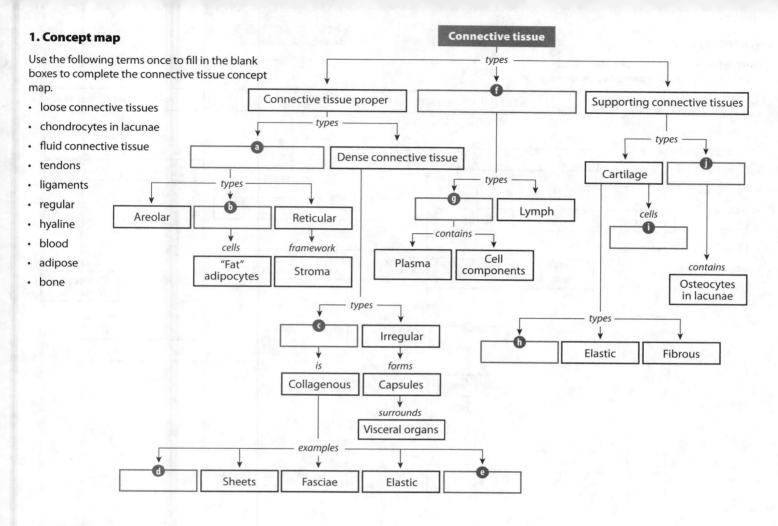

2. Vocabulary

In the space provided, write the boldfaced terms introduced in this section that contain the indicated word part.

Word Part	Meaning	Terms
peri-	around	**a** _____
os-	bone	**b** _____
chondro-	cartilage	**c** _____
inter-	between	**d** _____
lacus-	lake	**e** _____

Fill in the blanks with the appropriate term.

f _____ A cartilage cell
g _____ Bone tissue
h _____ A type of cartilage that has a matrix with little ground substance and large amounts of collagen fibers
i _____ Cells that store lipid reserves
j _____ The membrane that lines mobile joint cavities
k _____ The membrane that covers the surface of the body
l _____ Separates cartilage from surrounding tissues

1. Concept map

Use the following terms once to fill in the blank boxes to complete the connective tissue concept map.

- loose connective tissues
- chondrocytes in lacunae
- fluid connective tissue
- tendons
- ligaments
- regular
- hyaline
- blood
- adipose
- bone

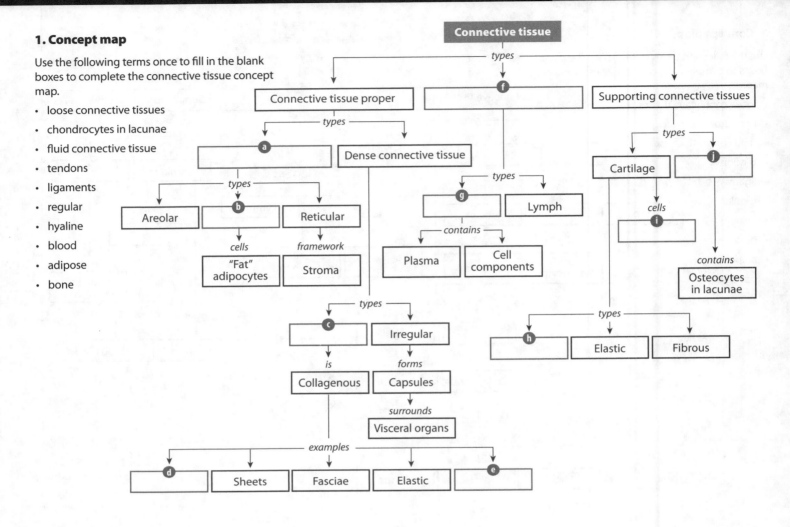

2. Vocabulary

In the space provided, write the boldfaced terms introduced in this section that contain the indicated word part.

Word Part	Meaning	Terms
peri-	around	**a** _____
os-	bone	**b** _____
chondro-	cartilage	**c** _____
inter-	between	**d** _____
lacus-	lake	**e** _____

Fill in the blanks with the appropriate term.

f _____ A cartilage cell

g _____ Bone tissue

h _____ A type of cartilage that has a matrix with little ground substance and large amounts of collagen fibers

i _____ Cells that store lipid reserves

j _____ The membrane that lines mobile joint cavities

k _____ The membrane that covers the surface of the body

l _____ Separates cartilage from surrounding tissues

1. Concept map

Fill in the blank boxes with boldfaced terms and concepts introduced in this section to complete the muscle tissue concept map.

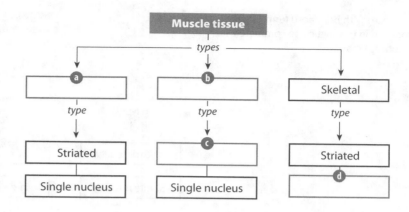

2. Concept map

Fill in the blank boxes with the boldfaced terms and concepts introduced in this section to complete the neural tissue concept map.

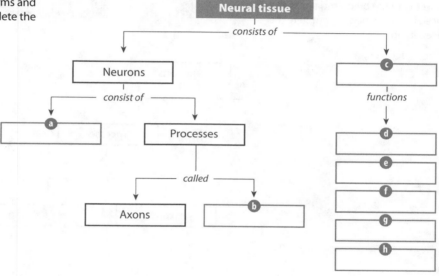

3. Vocabulary

Fill in the blanks with the appropriate term.

a _____ A single process that extends from the cell body of a neuron and carries information to other cells

b _____ A specialized intercellular junction between cardiac muscle cells

c _____ The supporting cells found in nervous tissue

d _____ Muscle tissue that contains large, multinucleate, striated cells

e _____ Muscle tissue that regulates the diameter of blood vessels and respiratory passageways

f _____ The repair process that occurs after inflammation has subsided

g _____ The first process in a tissue's response to injury

4. Section integration

During inflammation, both blood flow and blood vessel permeability increase in the injured area.
Describe how these responses aid the cleanup process and eliminate the inflammatory stimuli in the injured area.

_____

_____

_____

1. Concept map

Fill in the blank boxes with boldfaced terms and concepts introduced in this section to complete the muscle tissue concept map.

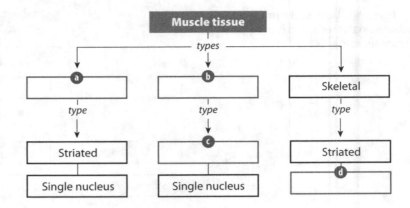

2. Concept map

Fill in the blank boxes with the boldfaced terms and concepts introduced in this section to complete the neural tissue concept map.

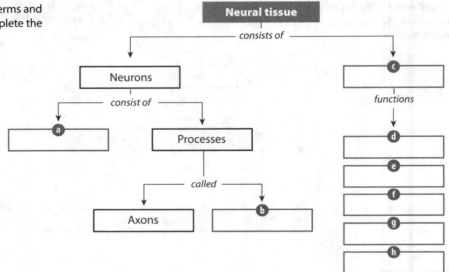

3. Vocabulary

Fill in the blanks with the appropriate term.

a _____ A single process that extends from the cell body of a neuron and carries information to other cells

b _____ A specialized intercellular junction between cardiac muscle cells

c _____ The supporting cells found in nervous tissue

d _____ Muscle tissue that contains large, multinucleate, striated cells

e _____ Muscle tissue that regulates the diameter of blood vessels and respiratory passageways

f _____ The repair process that occurs after inflammation has subsided

g _____ The first process in a tissue's response to injury

4. Section integration

During inflammation, both blood flow and blood vessel permeability increase in the injured area.
Describe how these responses aid the cleanup process and eliminate the inflammatory stimuli in the injured area.

_____

_____

_____

Visual Outline with Key Terms

Summarize the content of each module using the terms in the order provided.

SECTION 1

Epithelial Tissue

- tissues
- histology

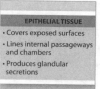

EPITHELIAL TISSUE
- Covers exposed surfaces
- Lines internal passageways and chambers
- Produces glandular secretions

4.1

Epithelial tissue covers surfaces, lines structures, and forms secretory glands

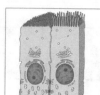

- epithelia
- glands
- exocrine glands
- endocrine glands
- apical surface
- base
- polarity
- lumen
- basolateral surfaces
- squamous
- cuboidal
- columnar
- simple epithelium
- stratified epithelium

4.2

Epithelial cells are extensively interconnected, both structurally and functionally

- occluding junction
- adhesion belt
- gap junction
- desmosomes
- hemidesmosomes
- basal lamina
- clear layer
- dense layer
- connexons
- cell adhesion molecules (CAMs)
- intercellular cement
- hyaluronan
- avascular

• = *Term boldfaced in this module*

4.3

The cells in a squamous epithelium are flattened and irregular in shape

- squamous epithelium
- simple squamous epithelium
- mesothelium
- endothelium
- stratified squamous epithelium
- keratininzed
- nonkeratinized

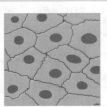

4.4

Cuboidal and transitional epithelia are found along several passageways and chambers connected to the exterior

- cuboidal epithelium
- simple cuboidal epithelium
- stratified cuboidal epithelium
- transitional epithelium

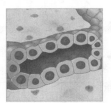

4.5

Columnar epithelia typically perform absorption or provide protection from chemical or environmental stresses

- columnar epithelium
- simple columnar epithelium
- pseudostratified columnar epithelium
- stratified columnar epithelium

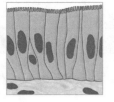

• = *Term boldfaced in this module*

4.6

Glandular epithelia are specialized for secretion

- glands
- endocrine glands
- exocrine glands
- merocrine secretion
- mucin
- mucus
- apocrine secretion
- holocrine secretion
- simple gland
- branched gland
- tubular gland
- alveolar (acinar) gland
- compound gland
- tubuloalveolar gland
- mucous cells

SECTION 2

Connective Tissue

- ground substance
- matrix
- connective tissue proper
- fluid connective tissues
- supporting connective tissues

4.7

Loose connective tissues provide padding and a supporting framework for other tissue types

- loose connective tissues
- areolar tissue
- reticular fibers
- collagen fibers
- elastic fibers
- melanocyte
- fixed macrophage
- mast cells
- fibroblasts
- adipocytes
- plasma cell
- free macrophages
- mesenchymal cells
- neutrophils
- eosinophils
- lymphocytes
- adipose tissue
- reticular tissue
- stroma

● = *Term boldfaced in this module*

4.8

Dense connective tissues are dominated by extracellular fibers, whereas fluid connective tissues have a liquid matrix

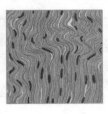

- dense connective tissues
- dense regular connective tissue
- dense irregular connective tissue
- elastic tissue
- fluid connective tissues
- plasma
- formed elements
- red blood cells
- white blood cells
- monocytes
- lymphocytes
- eosinophils
- neutrophils
- basophils
- platelets
- blood
- lymph
- arteries
- capillaries
- veins

4.9

Cartilage provides a flexible supporting framework

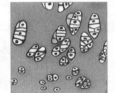

- cartilage
- chondroitin sulfates
- chondrocytes
- lacunae
- hyaline cartilage
- elastic cartilage
- fibrous cartilage
- perichondrium
- appositional growth
- chondroblasts
- interstitial growth

4.10

Bone provides a strong framework for the body

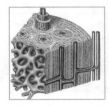

- osseous tissue
- compact bone
- spongy bone
- periosteum
- osteocytes
- canaliculi
- central canal
- osteon

• = *Term boldfaced in this module*

4.11

Membranes are physical barriers, and fasciae create internal compartments and divisions

- mucous membranes
- lamina propria
- serous membranes
- pleura
- peritoneum
- pericardium
- transudate
- cutaneous membrane
- synovial membrane
- synovial fluid
- fasciae
- superficial fascia
- deep fascia
- subserous fascia

_____

_____

_____

_____

_____

_____

_____

_____

_____

_____

_____

_____

_____

_____

_____

_____

_____

_____

SECTION 3

Muscle Tissue and Neural Tissue

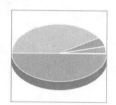

- muscle tissue
- neural tissue

_____

_____

_____

_____

_____

_____

_____

_____

_____

_____

_____

_____

_____

_____

4.12

Muscle tissue is specialized for contraction; neural tissue is specialized for communication

- muscle tissue
- skeletal muscle tissue
- striated
- multinucleate
- cardiac muscle tissue
- intercalated discs
- smooth muscle tissue
- neural tissue
- neurons
- neuroglia (glial cells)
- dendrites
- axon
- cell body

_____

_____

_____

_____

_____

_____

_____

_____

_____

_____

_____

_____

_____

_____

_____

• = _Term boldfaced in this module_

4.13

The response to tissue injury involves inflammation and regeneration

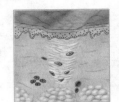

- mast cell activation
- inflammation
- infection
- dilate

- regeneration
- scar tissue
- fibrosis

_____

_____

_____

_____

_____

_____

_____

_____

_____

_____

1. Short answer

Identify and describe the layers of the cutaneous membrane and the underlying layer of loose connective tissue in the diagram at right.

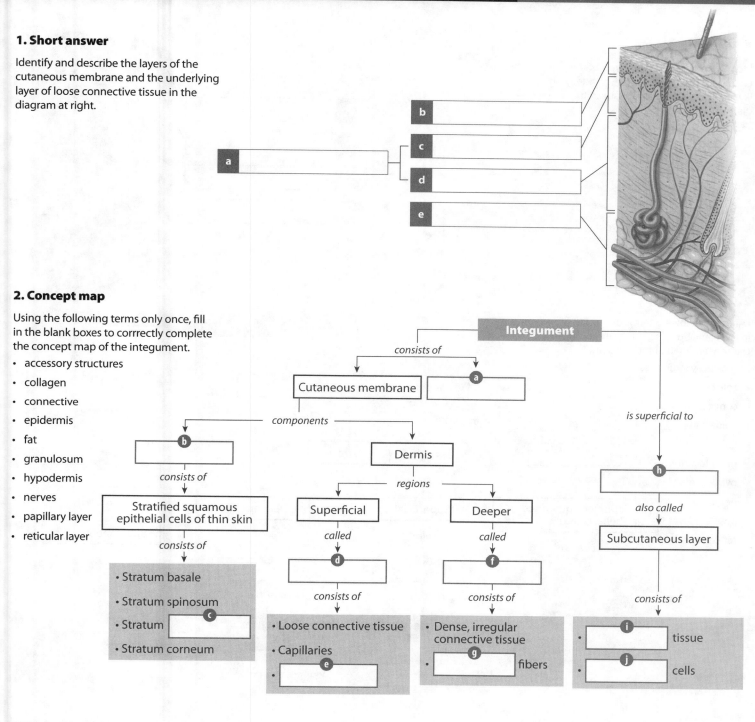

2. Concept map

Using the following terms only once, fill in the blank boxes to corrrectly complete the concept map of the integument.

- accessory structures
- collagen
- connective
- epidermis
- fat
- granulosum
- hypodermis
- nerves
- papillary layer
- reticular layer

Integument

consists of

Cutaneous membrane | a

components

b

consists of

Stratified squamous epithelial cells of thin skin

consists of

- Stratum basale
- Stratum spinosum
- Stratum c
- Stratum corneum

Dermis

regions

Superficial | Deeper

called | *called*

d | f

consists of | *consists of*

- Loose connective tissue
- Capillaries
- e

- Dense, irregular connective tissue
- g fibers

is superficial to

h

also called

Subcutaneous layer

consists of

- i tissue
- j cells

3. System integration

a. Describe why malignancies of melanocytes are so often fatal.

_____

_____

b. After taking a long bubble bath or doing dishes by hand, your fingers often appear wrinkled, or "pruny." Explain why.

_____

_____

1. Short answer

Identify and describe the layers of the cutaneous membrane and the underlying layer of loose connective tissue in the diagram at right.

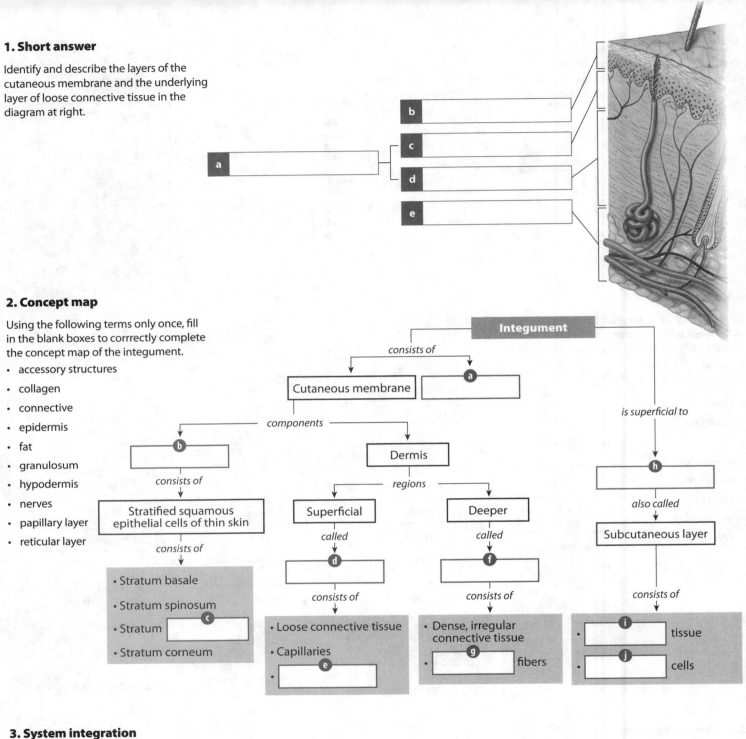

2. Concept map

Using the following terms only once, fill in the blank boxes to corrrectly complete the concept map of the integument.

- accessory structures
- collagen
- connective
- epidermis
- fat
- granulosum
- hypodermis
- nerves
- papillary layer
- reticular layer

Integument

consists of

Cutaneous membrane   **a**

components

b

consists of

Stratified squamous epithelial cells of thin skin

consists of

- Stratum basale
- Stratum spinosum
- Stratum **c**
- Stratum corneum

Dermis

regions

Superficial

called

d

consists of

- Loose connective tissue
- Capillaries
- **e**

Deeper

called

f

consists of

- Dense, irregular connective tissue
- **g** fibers

is superficial to

h

also called

Subcutaneous layer

consists of

- **i** tissue
- **j** cells

3. System integration

a. Describe why malignancies of melanocytes are so often fatal.

_____

_____

b. After taking a long bubble bath or doing dishes by hand, your fingers often appear wrinkled, or "pruny." Explain why.

_____

_____

1. Matching

Match the following terms with the most closely related description.

- malignant melanoma
- wrinkled skin
- nail root
- sebum
- apocrine sweat glands
- eponychium
- EGF
- vitamin D₃
- reticular layer of dermis
- merocrine sweat glands

a _____ Produced by epidermal cells stimulated by UV radiation

b _____ Epithelial fold not visible from the surface

c _____ Found in the armpit

d _____ Peptide produced by salivary glands

e _____ Site of hair production

f _____ Decrease in elastic fibers

g _____ Oily lipid secretion

h _____ Melanocytes metastasize through the lymphatic system

i _____ Abundant in the palms and soles

j _____ Cuticle

2. Labeling

Label the structures of a typical nail in the accompanying figures.

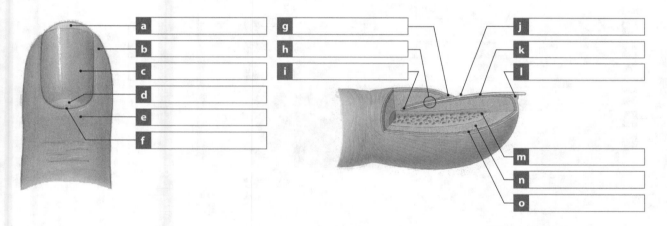

3. Labeling

Label the structures of a hair follicle in the accompanying figure.

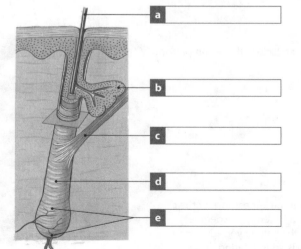

4. Section integration

Many people change the natural appearance of their hair, either by coloring it or by altering the degree of curl in it. Which layers of the hair do you suppose are affected by the chemicals added during these procedures? Why are the effects of the procedures not permanent?

1. Matching

Match the following terms with the most closely related description.

- malignant melanoma
- wrinkled skin
- nail root
- sebum
- apocrine sweat glands
- eponychium
- EGF
- vitamin D_3
- reticular layer of dermis
- merocrine sweat glands

a	_____	Produced by epidermal cells stimulated by UV radiation
b	_____	Epithelial fold not visible from the surface
c	_____	Found in the armpit
d	_____	Peptide produced by salivary glands
e	_____	Site of hair production
f	_____	Decrease in elastic fibers
g	_____	Oily lipid secretion
h	_____	Melanocytes metastasize through the lymphatic system
i	_____	Abundant in the palms and soles
j	_____	Cuticle

2. Labeling

Label the structures of a typical nail in the accompanying figures.

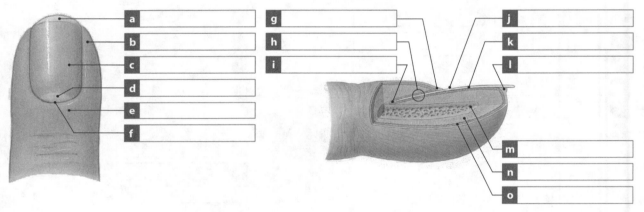

3. Labeling

Label the structures of a hair follicle in the accompanying figure.

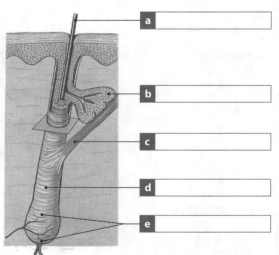

4. Section integration

Many people change the natural appearance of their hair, either by coloring it or by altering the degree of curl in it. Which layers of the hair do you suppose are affected by the chemicals added during these procedures? Why are the effects of the procedures not permanent?

Visual Outline with Key Terms

Summarize the content of each module using the terms in the order provided.

SECTION 1

Functional Anatomy of the Skin

- integumentary system
- integument
- cutaneous membrane
- epidermis
- dermis
- hypodermis
- accessory structures
- cutaneous plexus

_____

_____

_____

_____

_____

_____

_____

_____

5.1

The epidermis is composed of layers with various functions

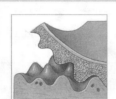

- keratinocytes
- epidermal ridges
- dermal papillae
- thin skin
- thick skin
- stratum corneum
- keratin
- stratum lucidum
- keratohyalin
- stratum granulosum
- stratum spinosum
- dendritic cells
- stratum basale
- basal cells
- Merkel cells

_____

_____

_____

_____

_____

_____

_____

_____

_____

5.2

Factors influencing skin color include epidermal pigmentation and dermal circulation

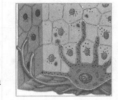

- melanin
- melanocyte
- melanosomes
- carotene
- cyanosis
- papillary plexus
- basal cell carcinoma
- malignant melanoma

_____

_____

_____

_____

_____

_____

_____

● = *Term boldfaced in this module*

5.3

The dermis supports the epidermis, and the subcutaneous layer connects the dermis to the rest of the body

- dermis
- papillary layer
- reticular layer
- hypodermis
- tactile discs
- tactile corpuscles
- lamellated corpuscles
- lines of cleavage

SECTION 2

Accessory Organs of the Skin

- hair follicles
- exocrine glands
- nails

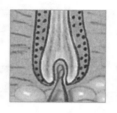

5.4

Hair is composed of dead, keratinized cells produced in a specialized hair follicle

- hairs
- hair follicle
- terminal hairs
- vellus hairs
- hair shaft
- hair root
- root hair plexus
- sebaceous gland
- arrector pili
- hair bulb
- hair papilla
- cuticle
- cortex
- medulla
- hair matrix
- soft keratin
- hard keratin
- internal root sheath
- external root sheath
- glassy membrane
- hair growth cycle
- active phase
- resting phase
- club hair

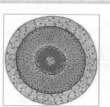

• = *Term boldfaced in this module*

5.5

Sebaceous glands and sweat glands are exocrine glands found in the skin

- sebaceous glands
- sebum
- sebaceous follicles
- sweat glands
- apocrine sweat glands
- merocrine sweat glands
- myoepithelial cells

5.6

Nails are thick sheets of keratinized epidermal cells that protect the tips of fingers and toes

- nails
- nail body
- lateral nail grooves
- lateral nail folds
- nail bed
- lunula
- nail root
- eponychium
- hyponychium

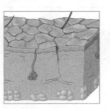

5.7

Age-related changes alter the integument

- fewer melanocytes
- drier epidermis
- thinning epidermis
- diminished immune response
- thinning dermis
- decreased perspiration
- altered hair and fat distribution
- fewer active hair follicles
- slower skin repair
- reduced blood supply

• = *Term boldfaced in this module*

5.8

The integument both responds to circulating hormones and has endocrine functions that are dependent on ultraviolet radiation

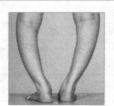

- glucocorticoids
- thyroid gland hormones
- sex hormones
- growth factors
- epidermal growth factor (EGF)
- growth hormone (GH)
- cholecalciferol (vitamin D$_3$)
- sunlight
- diet
- calcitriol
- rickets

5.9

The integument can often repair itself, even after extensive damage

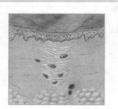

- scab
- granulation tissue
- scar tissue
- keloid

• = *Term boldfaced in this module*

1. Vocabulary

In the space provided, write the term for each of the following definitions.

a _____ Bones with complex shapes

b _____ The expanded ends of a long bone

c _____ A shallow depression in the surface of a bone

d _____ The marrow-filled space within a bone

e _____ The strut- and plate-shaped matrix of spongy bone

f _____ Cells that remove and recycle bone matrix

g _____ Bones that develop in tendons

h _____ The process that forms new bone matrix

i _____ The basic functional unit of compact bone

j _____ Type of bone growth that increases bone diameter

k _____ Process by which cartilage is replaced by bone

2. Concept map

Use each of the following terms once to fill in the blank boxes to correctly complete the bone formation concept map.

- lacunae
- osteocytes
- collagen
- intramembranous ossification
- compact bone
- periosteum
- hyaline cartilage

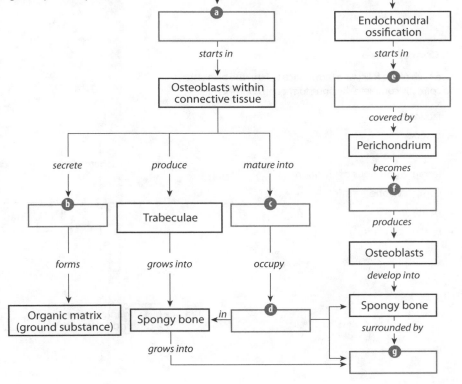

3. Section integration

While playing on her swing set, 10-year-old Rebecca falls and breaks her right leg. At the emergency room, the doctor tells her parents that the proximal end of the tibia where the epiphysis meets the diaphysis is fractured. The fracture is properly set and eventually heals. During a routine physical when she is 18, Rebecca learns that her right leg is 2 cm shorter than her left. What might account for this difference? _____

_____

_____

_____

1. Vocabulary

In the space provided, write the term for each of the following definitions.

a _____ Bones with complex shapes

b _____ The expanded ends of a long bone

c _____ A shallow depression in the surface of a bone

d _____ The marrow-filled space within a bone

e _____ The strut- and plate-shaped matrix of spongy bone

f _____ Cells that remove and recycle bone matrix

g _____ Bones that develop in tendons

h _____ The process that forms new bone matrix

i _____ The basic functional unit of compact bone

j _____ Type of bone growth that increases bone diameter

k _____ Process by which cartilage is replaced by bone

2. Concept map

Use each of the following terms once to fill in the blank boxes to correctly complete the bone formation concept map.

- lacunae
- osteocytes
- collagen
- intramembranous ossification
- compact bone
- periosteum
- hyaline cartilage

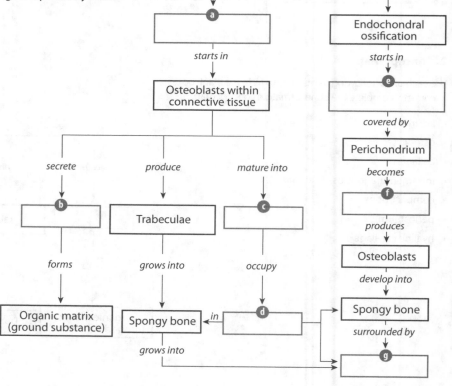

3. Section integration

While playing on her swing set, 10-year-old Rebecca falls and breaks her right leg. At the emergency room, the doctor tells her parents that the proximal end of the tibia where the epiphysis meets the diaphysis is fractured. The fracture is properly set and eventually heals. During a routine physical when she is 18, Rebecca learns that her right leg is 2 cm shorter than her left. What might account for this difference? _____

_____

_____

_____

1. Concept map

Use each of the following terms once to fill in the blank boxes to correctly complete the regulation of calcium ion concentration concept map.

- . $\downarrow$ Ca^{2+} level
- homeostasis
- release of stored Ca^{2+} from bone
- . $\downarrow$ Ca^{2+} concentration in body fluids
- parathyroid glands
- calcitonin
- . $\uparrow$ Ca^{2+} concentration in body fluids

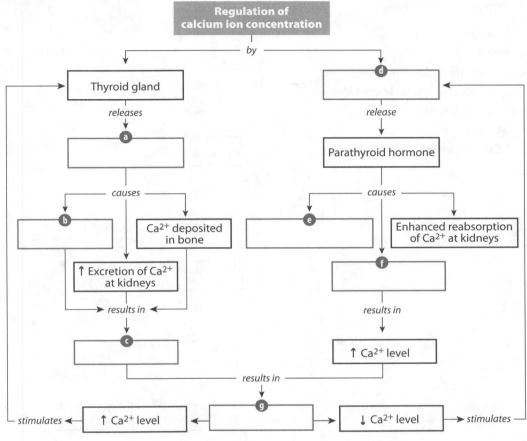

Regulation of calcium ion concentration

by

Thyroid gland

releases

a

causes

b | Ca^{2+} deposited in bone

$\uparrow$ Excretion of Ca^{2+} at kidneys

results in

c

d

release

Parathyroid hormone

causes

e | Enhanced reabsorption of Ca^{2+} at kidneys

f

results in

$\uparrow$ Ca^{2+} level

results in

stimulates ← $\uparrow$ Ca^{2+} level ← **g** → $\downarrow$ Ca^{2+} level → *stimulates*

2. Short answer

Identify the type of fracture and the bones involved in each of the following x-ray images. Circle the fracture that typically results from cushioning a fall.

a

b

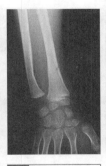

c

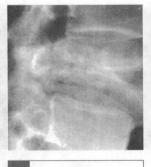

d

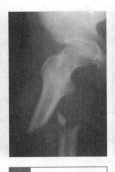

e

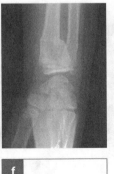

f

1. Concept map

Use each of the following terms once to fill in the blank boxes to correctly complete the regulation of calcium ion concentration concept map.

- $\downarrow Ca^{2+}$ level
- homeostasis
- release of stored Ca^{2+} from bone
- $\downarrow Ca^{2+}$ concentration in body fluids
- parathyroid glands
- calcitonin
- $\uparrow Ca^{2+}$ concentration in body fluids

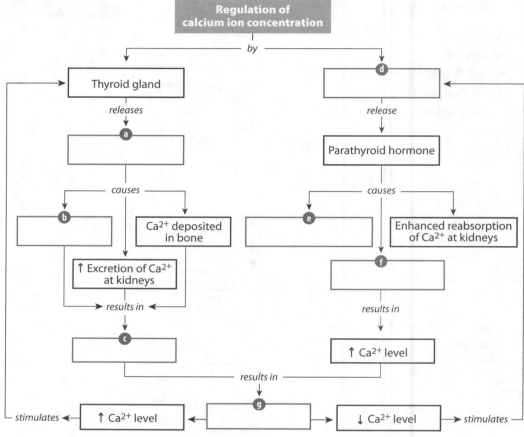

Regulation of calcium ion concentration

by

Thyroid gland

releases

a

causes

b Ca^{2+} deposited in bone

$\uparrow$ Excretion of Ca^{2+} at kidneys

results in

c

d

release

Parathyroid hormone

causes

e   Enhanced reabsorption of Ca^{2+} at kidneys

f

results in

$\uparrow Ca^{2+}$ level

results in

g

stimulates ← $\uparrow Ca^{2+}$ level ←   → $\downarrow Ca^{2+}$ level → *stimulates*

2. Short answer

Identify the type of fracture and the bones involved in each of the following x-ray images. Circle the fracture that typically results from cushioning a fall.

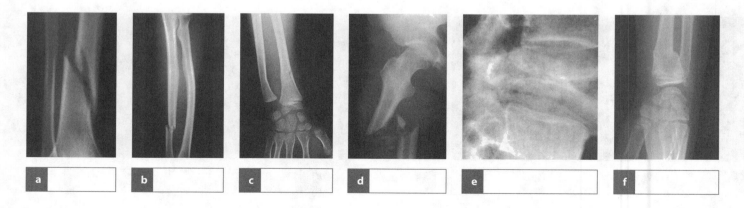

a

b

c

d

e

f

Visual Outline with Key Terms

Summarize the content of each module using the terms in the order provided.

SECTION 1

An Introduction to the Bones of the Skeletal System

- axial skeleton
- appendicular skeleton

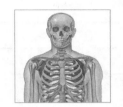

6.1

Bones are classified according to shape and structure, and display surface features

- flat bones
- sutural bones
- long bones
- irregular bones
- sesamoid bones
- short bones
- surface features
- canal (meatus)
- process
- sinus
- foramen

- fissure
- head
- tubercle
- sulcus
- tuberosity
- diaphysis
- trochlea
- condyle
- trochanter
- neck
- facet

- crest
- fossa
- line
- spine
- ramus

6.2

Long bones are designed to transmit forces along the shaft and have a rich blood supply

- epiphysis
- metaphysis
- diaphysis
- spongy (trabecular) bone
- compact bone
- medullary cavity
- red bone marrow
- yellow bone marrow

- nutrient artery
- nutrient vein
- nutrient foramen
- articular cartilage
- metaphyseal artery
- metaphyseal vein

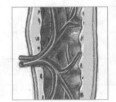

• = *Term boldfaced in this module*

6.3

Bone has a calcified matrix associated with osteocytes, osteoblasts, osteoprogenitor cells, and osteoclasts

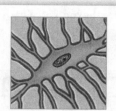

- osteocytes
- lacuna
- canaliculi
- lamellae
- osteoblasts
- osteogenesis
- ossification
- osteoid
- osteoprogenitor cells
- osteoclasts
- osteolysis
- hydroxyapatite

6.4

Compact bone consists of parallel osteons, and spongy bone consists of a network of trabeculae

- osteon
- concentric lamellae
- central canal
- circumferential lamellae
- interstitial lamellae
- perforating canals
- trabeculae

6.5

In appositional bone growth, layers of compact bone are added to the bone's outer surface

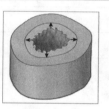

- appositional growth
- periosteum
- perforating fibers
- endosteum

• = *Term boldfaced in this module*

6.6

Endochondral ossification is the replacement of a cartilaginous model with bone

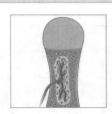

- endochondral ossification
- primary ossification center
- secondary ossification center
- epiphyseal cartilage
- epiphyseal line

6.7

Intramembranous ossification is the formation of bone without a prior cartilaginous model

- intramembranous ossification
- dermal bones
- ossification center
- spicules

6.8

Abnormalities of bone growth and development produce recognizable physical signs

- pituitary growth failure
- achondroplasia
- Marfan syndrome
- gigantism
- fibrodysplasia ossificans progressiva (FOP)
- heterotopic (ectopic) bones
- acromegaly

● = *Term boldfaced in this module*

The Physiology of Bones

- ○ composition of bone
- ○ homeostatic regulation of calcium
- ○ osteoblasts
- ○ osteoclasts
- • calcium

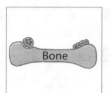

Bone

6.9

The primary hormones regulating calcium ion metabolism are parathyroid hormone and calcitonin; calcitriol is also involved

- • parathyroid glands
- • parathyroid hormone (PTH)
- ○ kidneys
- • calcitriol
- • thyroid gland
- • C cells
- • calcitonin

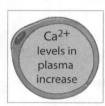

Ca²⁺ levels in plasma increase

6.10

A fracture is a crack or a break in a bone

- • fracture
- • fracture hematoma
- • internal callus
- • external callus
- • closed (simple) fractures
- • open (compound) fractures
- • transverse fractures
- • spiral fractures
- • displaced fractures
- • nondisplaced fractures
- • compression fractures
- • greenstick fracture
- • comminuted fractures
- • epiphyseal fractures
- • Pott fracture
- • Colles fracture

• = *Term boldfaced in this module*

1. Concept map

Use each of the following terms once to fill in the blank boxes to correctly complete the skeleton concept map.

- floating
- temporal
- mandible
- axial
- hyoid
- sacral
- vertebral column
- lacrimal
- xiphoid process
- occipital
- sternum
- skull
- thoracic
- longitudinal
- lumbar

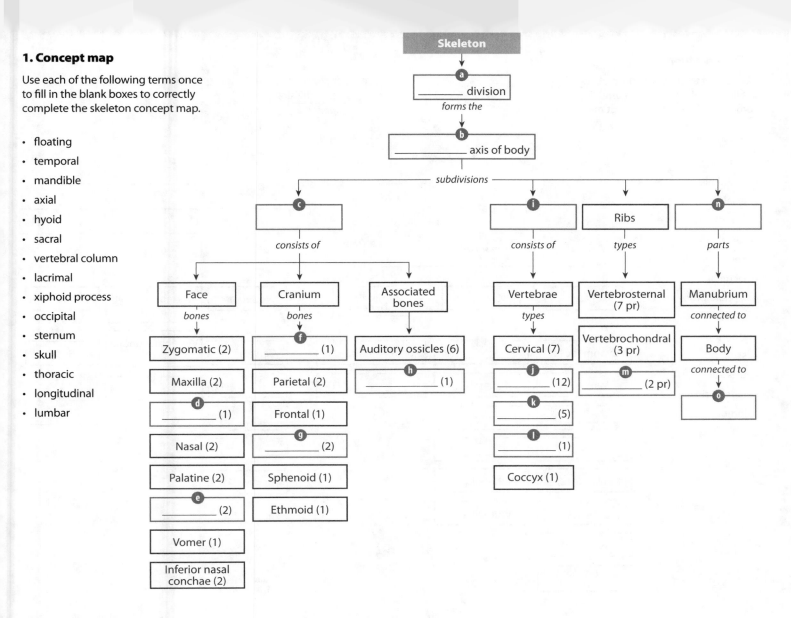

Skeleton

a _____ division

forms the

b _____ axis of body

subdivisions

c _____

consists of

Face	Cranium	Associated bones
bones	*bones*	
Zygomatic (2)	**f** _____ (1)	Auditory ossicles (6)
Maxilla (2)	Parietal (2)	**h** _____ (1)
d _____ (1)	Frontal (1)	
Nasal (2)	**g** _____ (2)	
Palatine (2)	Sphenoid (1)	
e _____ (2)	Ethmoid (1)	
Vomer (1)		
Inferior nasal conchae (2)		

i _____

consists of

Vertebrae

types

Cervical (7)
j _____ (12)
k _____ (5)
l _____ (1)
Coccyx (1)

Ribs

types

Vertebrosternal (7 pr)
Vertebrochondral (3 pr)
m _____ (2 pr)

n _____

parts

Manubrium

connected to

Body

connected to

o _____

2. Short answer

For each of the following vertebrae, identify its vertebral region.

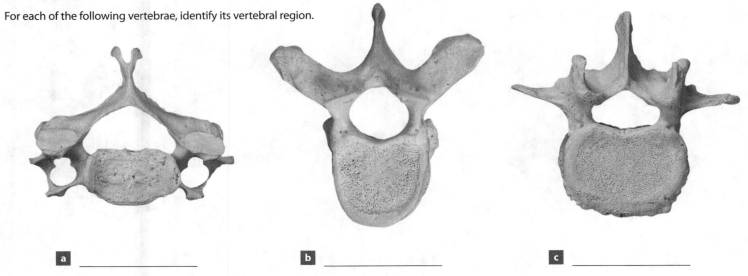

a _____

b _____

c _____

1. Concept map

Use each of the following terms once to fill in the blank boxes to correctly complete the skeleton concept map.

- floating
- temporal
- mandible
- axial
- hyoid
- sacral
- vertebral column
- lacrimal
- xiphoid process
- occipital
- sternum
- skull
- thoracic
- longitudinal
- lumbar

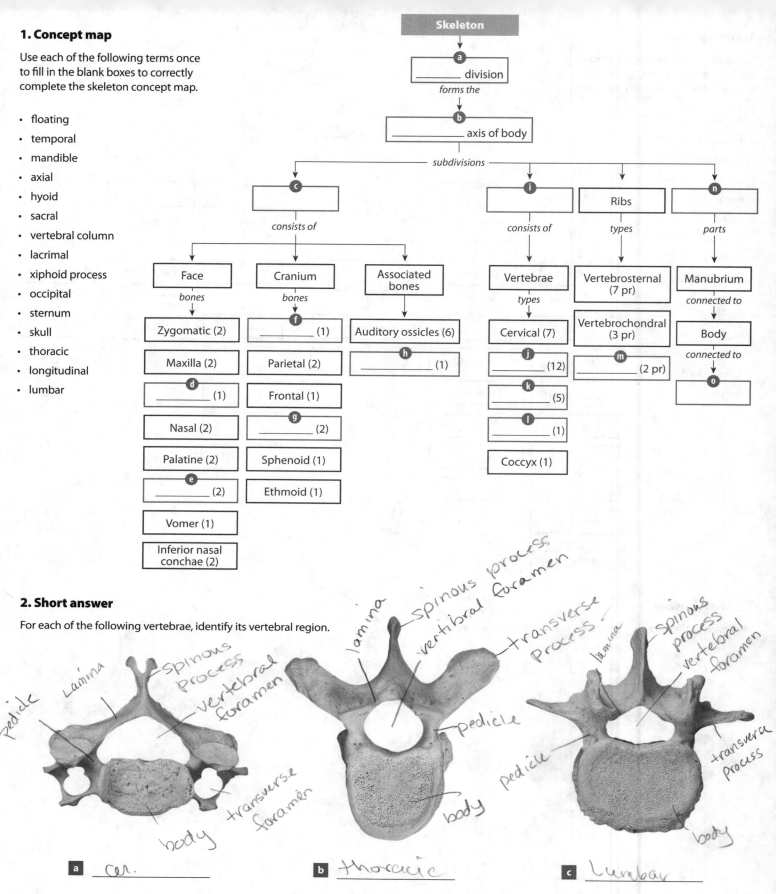

Skeleton

a _____ division

forms the

b _____ axis of body

subdivisions

c _____ — *consists of*

- Face — *bones*
  - Zygomatic (2)
  - Maxilla (2)
  - **d** _____ (1)
  - Nasal (2)
  - Palatine (2)
  - **e** _____ (2)
  - Vomer (1)
  - Inferior nasal conchae (2)
- Cranium — *bones*
  - **f** _____ (1)
  - Parietal (2)
  - Frontal (1)
  - **g** _____ (2)
  - Sphenoid (1)
  - Ethmoid (1)
- Associated bones
  - Auditory ossicles (6)
  - **h** _____ (1)

i _____ — *consists of*

- Vertebrae — *types*
  - Cervical (7)
  - **j** _____ (12)
  - **k** _____ (5)
  - **l** _____ (1)
  - Coccyx (1)

Ribs — *types*
- Vertebrosternal (7 pr)
- Vertebrochondral (3 pr)
- **m** _____ (2 pr)

n _____ — *parts*
- Manubrium — *connected to*
- Body — *connected to*
- **o** _____

2. Short answer

For each of the following vertebrae, identify its vertebral region.

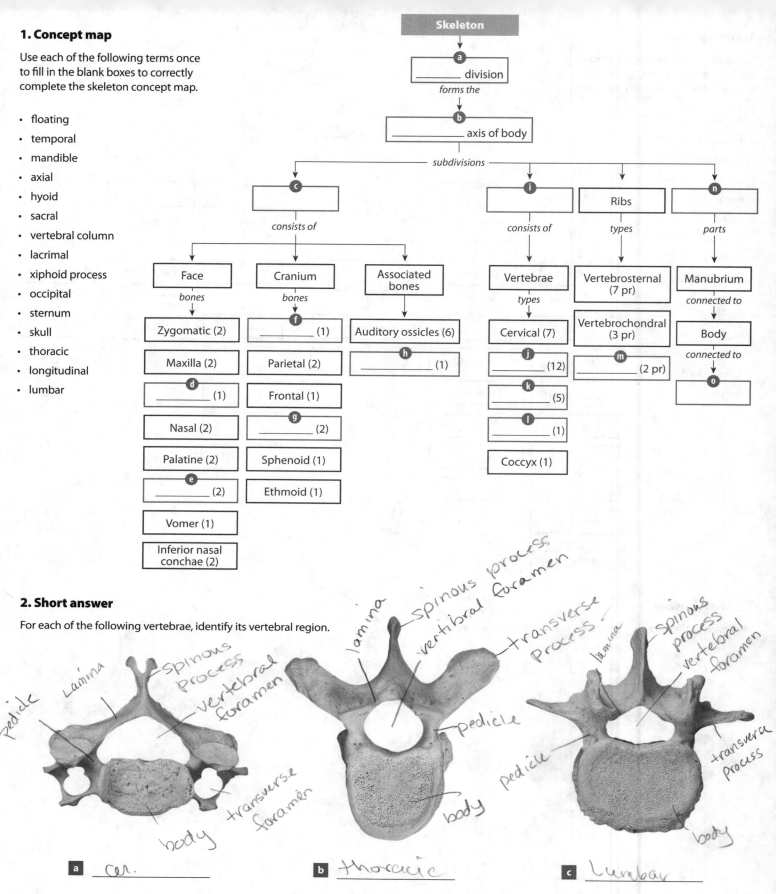

a cer.

b thoracic

c Lumbar

74

1. Labeling

Label the bones of the appendicular skeleton in the diagram at right.

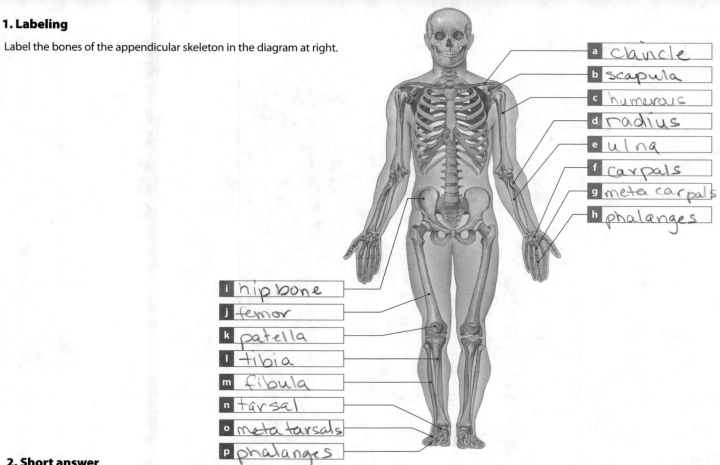

- a clavicle
- b scapula
- c humerus
- d radius
- e ulna
- f carpals
- g meta carpals
- h phalanges
- i hip bone
- j femor
- k patella
- l tibia
- m fibula
- n tarsal
- o meta tarsals
- p phalanges

2. Short answer

In the following photographs of a scapula, identify the three views (a–c) and the indicated bone markings (d–j).

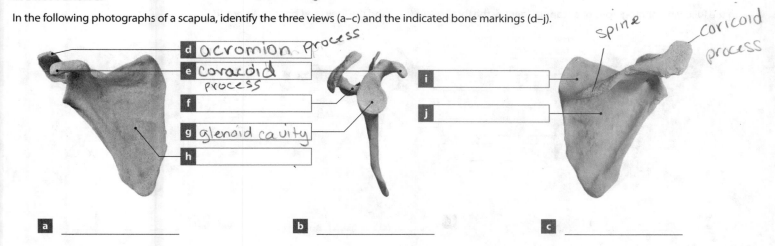

- d acromion process
- e coracoid process
- f
- g glenoid cavity
- h
- i
- j

spine

coricoid process

a _____

b _____

c _____

3. Labeling

In the following drawing of the pelvis, label the indicated structures.

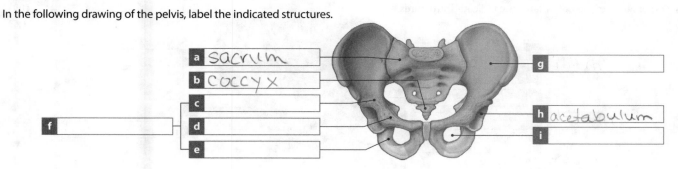

- a sacrium
- b coccyx
- c
- d
- e
- f
- g
- h acetabulum
- i

75

1. Labeling

Label the bones of the appendicular skeleton in the diagram at right.

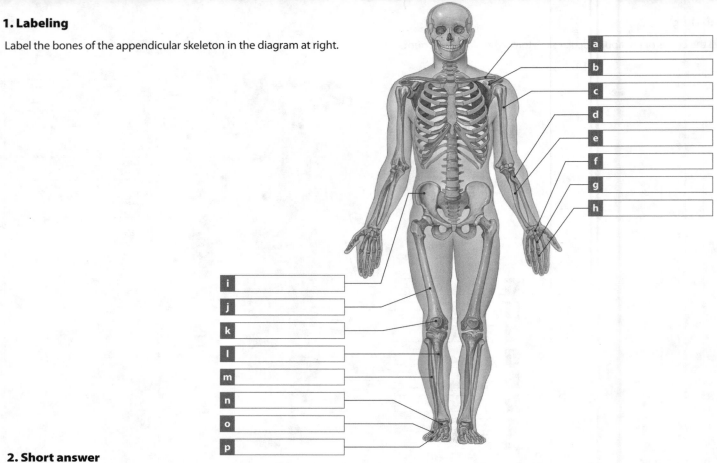

a _____

b _____

c _____

d _____

e _____

f _____

g _____

h _____

i _____

j _____

k _____

l _____

m _____

n _____

o _____

p _____

2. Short answer

In the following photographs of a scapula, identify the three views (a–c) and the indicated bone markings (d–j).

d _____

e _____

f _____

g _____

h _____

i _____

j _____

a _____

b _____

c _____

3. Labeling

In the following drawing of the pelvis, label the indicated structures.

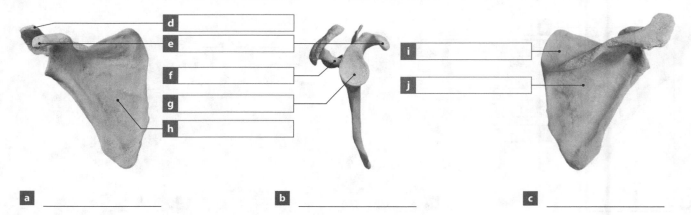

a _____

b _____

c _____

d _____

e _____

f _____

g _____

h _____

i _____

Visual Outline with Key Terms

Summarize the content of each module using the terms in the order provided.

SECTION 1

The Axial Skeleton

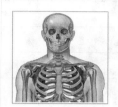

- axial skeleton
  - skull and associated bones
  - thoracic cage
  - vertebral column

_____

_____

_____

_____

_____

_____

_____

_____

_____

_____

_____

7.1

The skull has cranial and facial components that are usually bound together by sutures

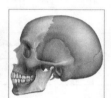

- facial bones
- cranial bones
- cranium
- cranial cavity
- sutures
- coronal suture
- calvaria
- squamous suture
- sagittal suture
- lambdoid suture

_____

_____

_____

_____

_____

_____

_____

_____

_____

_____

_____

_____

7.2

Facial bones dominate the anterior aspect of the skull, and cranial bones dominate the posterior surface

- nasal bones
- lacrimal bones
- palatine bones
- zygomatic bones
- maxillae
- inferior nasal conchae
- vomer
- mandible
- frontal bone
- sphenoid
- ethmoid
- parietal bones
- occipital bone
- temporal bones
- external occipital crest
- mastoid process
- styloid process

_____

_____

_____

_____

_____

_____

_____

_____

_____

_____

• = _Term boldfaced in this module_

7.3

The lateral and medial aspects of the skull share many surface features

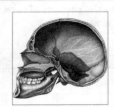

- frontal squama
- superior temporal lines
- inferior temporal lines
- squamous part
- external acoustic meatus
- zygomatic process
- zygomatic arch
- mandibular angle
- mental protuberance
- alveolar processes
- frontal sinuses
- hypophyseal fossa
- sella turcica
- petrous part
- internal acoustic meatus
- hypoglossal canal

7.4

The foramina on the inferior surface of the skull lead into the interior of the cranium

- mandibular fossa
- occipital condyles
- inferior and superior nuchal lines
- foramen lacerum
- foramen ovale
- carotid canal
- jugular foramen
- stylomastoid foramen
- foramen magnum
- internal occipital crest

7.5

The shapes and landmarks of the sphenoid, ethmoid, and palatine bones can best be seen in the isolated bones

- sphenoid
- optic canals
- lesser wings
- greater wings
- hypophyseal fossa
- sella turcica
- sphenoidal spine
- foramen spinosum
- foramen ovale
- foramen rotundum
- superior orbital fissure
- sphenoidal sinuses
- pterygoid plates
- body
- pterygoid processes
- ethmoid
- cribriform plate
- crista galli
- lateral masses
- ethmoidal labyrinth
- superior nasal conchae
- middle nasal conchae
- perpendicular plate (ethmoid)
- palatine bones
- perpendicular plate (palatine)
- horizontal plate
- orbital process

• = *Term boldfaced in this module*

7.6

Each orbital complex contains one eye, and the nasal complex encloses the nasal cavities

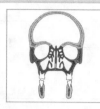

- orbital complex
- lacrimal fossa
- supra-orbital margin
- supra-orbital notch
- lacrimal sulcus
- nasolacrimal canal
- infra-orbital foramen
- zygomaticofacial foramen
- nasal complex
- paranasal sinuses
- ethmoidal air cells
- maxillary sinuses
- frontal sinuses
- sphenoidal sinuses

7.7

The mandible forms the lower jaw, and the associated bones of the skull perform specialized functions

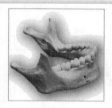

- condylar process
- mandibular notch
- coronoid process
- alveolar process
- body
- ramus
- mylohyoid line
- mandibular foramen
- hyoid bone
- greater horns
- lesser horns
- auditory ossicles

7.8

Fontanelles permit cranial growth in infants and small children

- fontanelles
- anterior fontanelle
- sphenoidal fontanelles
- mastoid fontanelles
- occipital fontanelle

• = Term boldfaced in this module

7.9

The vertebral column has four spinal curves, and vertebrae have both anatomical similarities and regional differences

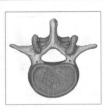

- cervical curve
- thoracic curve
- lumbar curve
- sacral curve
- articular processes
- vertebral arch
- vertebral body
- vertebral foramen
- spinous process
- laminae
- transverse processes
- pedicles
- intervertebral discs
- intervertebral foramina
- vertebral canal
- articular facet
- superior articular processes
- inferior articular processes

7.10

There are seven cervical vertebrae and twelve thoracic vertebrae

- cervical vertebrae
- transverse foramen
- bifid
- costal process
- atlas
- axis
- dens (odontoid process)
- vertebra prominens
- ligamentum nuchae
- thoracic vertebrae

7.11

There are five lumbar vertebrae; the sacrum and coccyx consist of fused vertebrae

- lumbar vertebrae
- sacrum
- ala
- sacral foramina
- base of sacrum
- sacral promontory
- apex of sacrum
- sacral canal
- superior articular processes
- median sacral crest
- sacral hiatus
- coccyx
- coccygeal cornua
- sacral tuberosity
- auricular surface
- lateral sacral crest

• = *Term boldfaced in this module*

7.12

The thoracic cage protects organs in the chest and provides sites for muscle attachment

- thoracic cage
- ribs
- vertebrosternal ribs
- costal cartilages
- vertebrochondral ribs
- floating ribs
- vertebral ribs
- jugular notch

- sternum
- manubrium
- body (of sternum)
- xiphoid process
- tubercle
- head (capitulum)
- articular facets
- angle of rib

- shaft of rib
- costal groove
- costal facets

SECTION 2

The Appendicular Skeleton

- appendicular skeleton
- pectoral girdle
- upper limbs
- pelvic girdles
- lower limbs

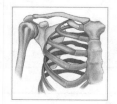

7.13

The pectoral girdles—the clavicles and scapulae—connect the upper limbs to the axial skeleton

- pectoral girdle
- clavicles
- scapulae
- sternal end
- acromion
- acromial end
  (of clavicle)
- body (of scapula)

- superior border
- medial border
- lateral border
- superior angle
- inferior angle
- lateral angle
- subscapular fossa
- glenoid cavity

- scapular spine
- supraspinous fossa
- infraspinous fossa
- coracoid process

• = *Term boldfaced in this module*

7.14

The humerus of the arm articulates with the radius and ulna of the forearm

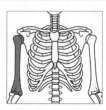

- humerus
- medial epicondyle
- lateral epicondyle
- intertubercular groove
- radial fossa
- capitulum
- condyle
- head (of humerus)
- lesser tubercle
- anatomical neck
- surgical neck
- deltoid tuberosity
- greater tubercle
- radial groove
- trochlea
- olecranon fossa
- coronoid fossa
- ulna
- radius
- olecranon
- ulnar head
- styloid process (ulna)
- radial head
- neck (of radius)
- radial tuberosity
- interosseous membrane
- ulnar notch
- styloid process (radius)
- trochlear notch
- coronoid process
- radial notch
- proximal radio-ulnar joint
- distal radio-ulnar joint

7.15

The wrist is composed of carpal bones, and the hand consists of metacarpal bones and phalanges

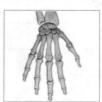

- carpus
- scaphoid
- lunate
- pisiform
- triquetrum
- trapezium
- trapezoid
- capitate
- hamate
- metacarpal bones
- phalanges
- pollex

7.16

The hip bone forms by the fusion of the ilium, ischium, and pubis

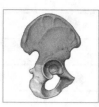

- pelvic girdle
- hip bones
- ilium
- ischium
- pubis
- iliac spines
- gluteal lines
- greater sciatic notch
- ischial spine
- ischial tuberosity
- iliac crest
- lunate surface
- acetabulum
- acetabular notch
- iliac fossa
- arcuate line
- pectineal line
- pubic tubercle
- pubic symphysis
- iliac tuberosity
- auricular surface (ilium)
- obturator foramen
- ischial ramus
- inferior pubic ramus
- superior pubic ramus

• = *Term boldfaced in this module*

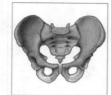

• = *Term boldfaced in this module*

1. Matching

Match the following terms with the most closely related description.

- amphiarthrosis
- synarthrosis
- dislocation
- pronation-supination
- diarthrosis
- shoulder
- articular discs
- fluid-filled pouch

a _____ Freely movable joint

b _____ Movements of forearm bones

c _____ Ball and socket

d _____ Menisci

e _____ Immovable joint

f _____ Luxation

g _____ Bursa

h _____ Slightly movable joint

2. Labeling

Label the structures in the synovial joint figure at right.

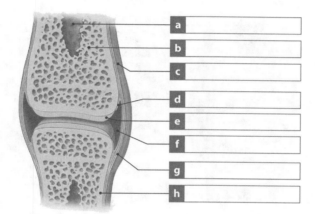

a _____

b _____

c _____

d _____

e _____

f _____

g _____

h _____

3. Labeling

Identify each of the following movements.

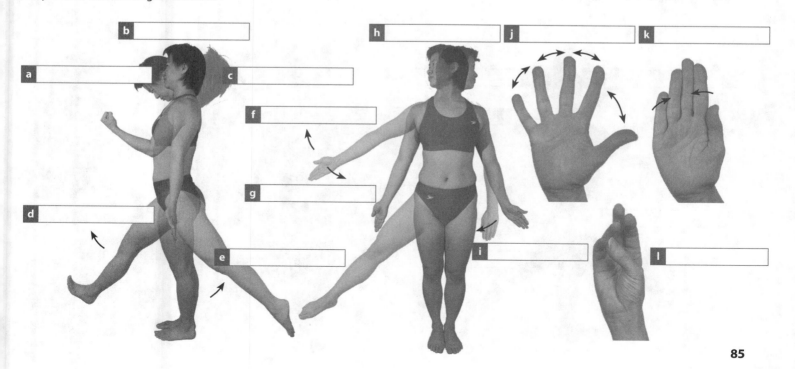

1. Matching

Match the following terms with the most closely related description.

- amphiarthrosis
- synarthrosis
- dislocation
- pronation-supination
- diarthrosis
- shoulder
- articular discs
- fluid-filled pouch

a _____ Freely movable joint

b _____ Movements of forearm bones

c _____ Ball and socket

d _____ Menisci

e _____ Immovable joint

f _____ Luxation

g _____ Bursa

h _____ Slightly movable joint

2. Labeling

Label the structures in the synovial joint figure at right.

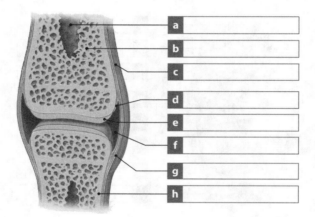

a _____

b _____

c _____

d _____

e _____

f _____

g _____

h _____

3. Labeling

Identify each of the following movements.

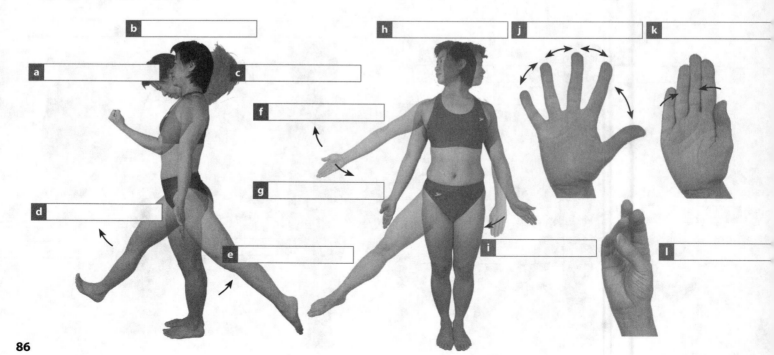

1. Matching

Match the following terms with the most closely related description.

- acetabulum
- popliteal ligament
- disc outer layer
- dislocation
- arthritis
- disc inner layer
- reinforce knee joint
- osteoporosis

a _____ Knee joint posterior

b _____ Articular cartilage damage

c _____ Reduced bone mass

d _____ Cruciate ligaments

e _____ Anulus fibrosus

f _____ Deep fossa

g _____ Nucleus pulposus

h _____ Nursemaid's elbow

2. Labeling

Label each of the structures in the accompanying diagram of the shoulder joint.

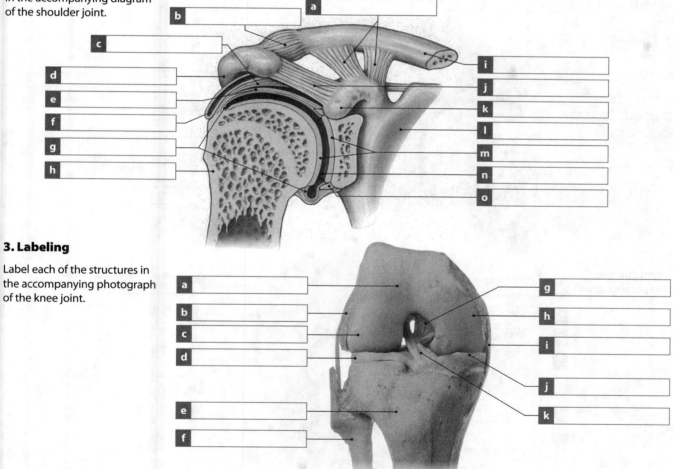

3. Labeling

Label each of the structures in the accompanying photograph of the knee joint.

4. Short answer

Identify the basic articulations between the pectoral and pelvic girdles (of the appendicular skeleton) and the axial skeleton.

_____

_____

1. Matching

Match the following terms with the most closely related description.

- acetabulum
- popliteal ligament
- disc outer layer
- dislocation
- arthritis
- disc inner layer
- reinforce knee joint
- osteoporosis

a	_____	Knee joint posterior
b	_____	Articular cartilage damage
c	_____	Reduced bone mass
d	_____	Cruciate ligaments
e	_____	Anulus fibrosus
f	_____	Deep fossa
g	_____	Nucleus pulposus
h	_____	Nursemaid's elbow

2. Labeling

Label each of the structures in the accompanying diagram of the shoulder joint.

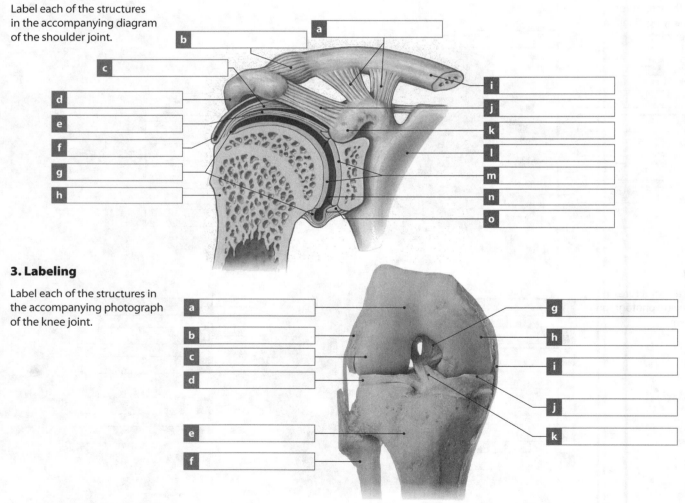

3. Labeling

Label each of the structures in the accompanying photograph of the knee joint.

4. Short answer

Identify the basic articulations between the pectoral and pelvic girdles (of the appendicular skeleton) and the axial skeleton.

_____

_____

Visual Outline with Key Terms

Summarize the content of each module using the terms in the order provided.

SECTION 1

Joint Design and Movement

- articulations
- range of motion (ROM)
- synarthrosis
- suture
- gomphosis
- synchondrosis
- synostosis
- amphiarthrosis
- syndesmosis
- symphysis
- diarthrosis
- diarthroses (synovial joints)

_____
_____
_____
_____
_____
_____
_____
_____
_____

8.1

Synovial joints are freely movable diarthroses containing synovial fluid

- articular cartilages
- joint capsule
- synovial fluid
- bursa
- fat pads
- meniscus
- accessory ligaments
- intrinsic ligaments
- extrinsic ligaments
- extracapsular ligaments
- intracapsular ligaments
- dislocation
- luxation

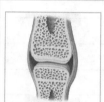

_____
_____
_____
_____
_____
_____
_____
_____
_____
_____
_____

8.2

Anatomical organization determines the functional properties of synovial joints

- gliding
- angular motion
- circumduction
- rotation
- gliding joint
- hinge joint
- pivot joint
- ellipsoid joint
- saddle joint
- ball-and-socket joint

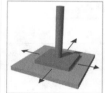

_____
_____
_____
_____
_____
_____
_____
_____
_____
_____

• = _Term boldfaced in this module_

8.3

Broad descriptive terms are used to describe movements with reference to the anatomical position

- flexion
- extension
- hyperextension
- lateral flexion
- dorsiflexion
- plantar flexion
- abduction
- adduction
- circumduction

8.4

Terms of more limited application describe rotational movements and special movements

- left rotation
- right rotation
- medial rotation
- lateral rotation
- pronation
- supination
- opposition
- protraction
- retraction
- inversion
- eversion
- depression
- elevation

SECTION 2

Articulations of the Axial and Appendicular Skeletons

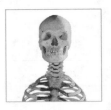

- joints of the axial skeleton
- joints of the appendicular skeleton

• = *Term boldfaced in this module*

8.5

Adjacent vertebrae have gliding diarthroses between their articular processes, and symphyseal joints between their vertebral bodies

- intervertebral discs
- anulus fibrosus
- nucleus pulposus
- ligamentum flavum
- posterior longitudinal ligament
- interspinous ligament
- supraspinous ligament

- anterior longitudinal ligament
- slipped disc
- herniated disc
- osteopenia
- osteoporosis

8.6

The shoulder and hip are ball-and-socket joints

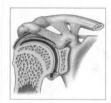

- glenohumeral joint
- glenoid labrum
- ligament of the femoral head
- transverse acetabular ligament
- acetabular labrum
- ligamentum teres

8.7

The elbow and knee are hinge joints

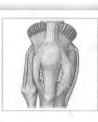

- humeroradial joint
- humero-ulnar joint
- nursemaid's elbow
- radial collateral ligament
- annular ligament
- ulnar collateral ligament
- fibular collateral ligament
- tibial collateral ligament
- popliteal ligaments

- anterior cruciate ligament (ACL)
- posterior cruciate ligament (PCL)
- medial menisci
- lateral menisci

• = *Term boldfaced in this module*

8.8

Arthritis can disrupt normal joint structure and function

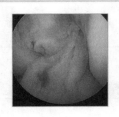

- rheumatism
- arthritis
- osteoarthritis
- arthroscope
- arthroscopic surgery

_____

_____

_____

_____

_____

_____

_____

1. Labeling

Label the structures in the following figure of a skeletal muscle fiber.

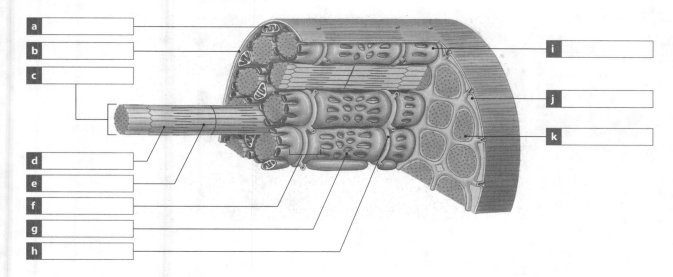

a _____
b _____
c _____
d _____
e _____
f _____
g _____
h _____
i _____
j _____
k _____

2. Labeling

Label the structures in the following diagram of adjacent sarcomeres.

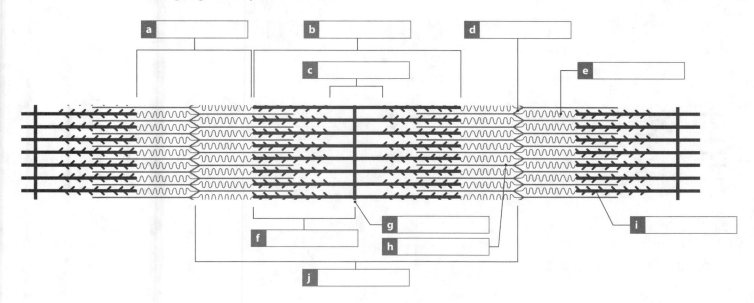

a _____
b _____
c _____
d _____
e _____
f _____
g _____
h _____
i _____
j _____

3. Vocabulary

In the space provided, write the boldfaced terms introduced in this section that contain the indicated word part.

Word Part	Meaning	Terms
myo-	muscle	a _____
sarko-	flesh	b _____

1. Labeling

Label the structures in the following figure of a skeletal muscle fiber.

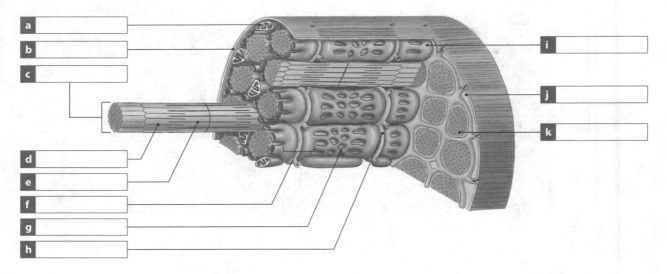

2. Labeling

Label the structures in the following diagram of adjacent sarcomeres.

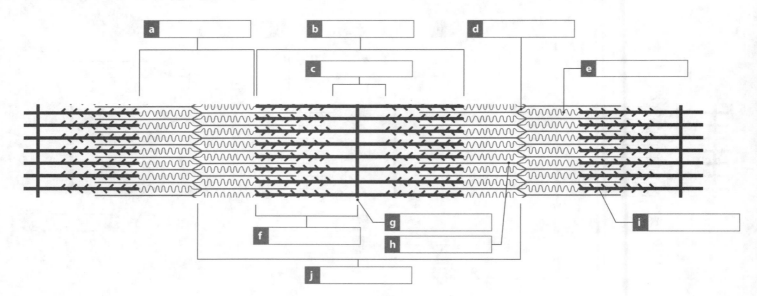

3. Vocabulary

In the space provided, write the boldfaced terms introduced in this section that contain the indicated word part.

Word Part	Meaning	Terms
myo-	muscle	**a**
sarko-	flesh	**b**

1. Matching

Match the following terms with the most closely related description.

- eccentric contraction
- isometric contraction
- isotonic contraction
- concentric contraction

a _____ Muscle does not change length during contraction

b _____ Muscle changes length during contraction

c _____ Peak tension less than load, and muscle elongates

d _____ Peak tension greater than load, and muscle shortens

2. Labeling

Use the following terms to correctly label the structures in the diagram representing a blood vessel and a skeletal muscle at rest.

- creatine
- glycogen
- CP
- glucose
- O_2
- fatty acids

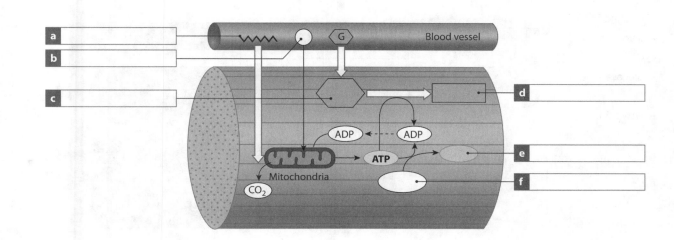

3. Short answer

Complete the following table by writing in the anatomical and physiological properties of the three types of skeletal muscle fibers.

Property	Slow Fibers	Intermediate Fibers	Fast Fibers
Cross-sectional diameter	a	b	c
Color	d	pink	e
Myoglobin content	high	f	g
Capillary supply	h	i	scarce
Mitochondria	j	k	l
Time to peak tension	m	medium	n
Contraction speed	o	p	q
Fatigue resistance	high	r	s
Glycolytic enzyme concentration in sarcoplasm	t	u	high

1. Matching

Match the following terms with the most closely related description.

- eccentric contraction
- isometric contraction
- isotonic contraction
- concentric contraction

a _____ Muscle does not change length during contraction

b _____ Muscle changes length during contraction

c _____ Peak tension less than load, and muscle elongates

d _____ Peak tension greater than load, and muscle shortens

2. Labeling

Use the following terms to correctly label the structures in the diagram representing a blood vessel and a skeletal muscle at rest.

- creatine
- glycogen
- CP
- glucose
- O_2
- fatty acids

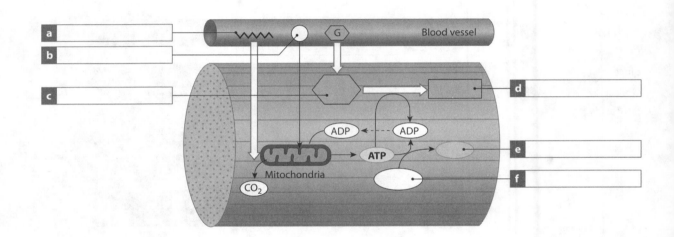

3. Short answer

Complete the following table by writing in the anatomical and physiological properties of the three types of skeletal muscle fibers.

Property	Slow Fibers	Intermediate Fibers	Fast Fibers
Cross-sectional diameter	a	b	c
Color	d	pink	e
Myoglobin content	high	f	g
Capillary supply	h	i	scarce
Mitochondria	j	k	l
Time to peak tension	m	medium	n
Contraction speed	o	p	q
Fatigue resistance	high	r	s
Glycolytic enzyme concentration in sarcoplasm	t	u	high

Visual Outline with Key Terms

Summarize the content of each module using the terms in the order provided.

SECTION 1

Functional Anatomy of Skeletal Muscle Tissue

- skeletal muscles
- skeletal muscle tissue
- cardiac muscle tissue
- smooth muscle tissue

9.1

A skeletal muscle contains skeletal muscle tissue, connective tissues, blood vessels, and nerves

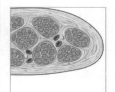

- skeletal muscle
- epimysium
- tendon
- aponeurosis
- muscle fascicle
- perimysium
- muscle fibers
- endomysium
- myosatellite cells
- myoblasts
- sarcolemma
- sarcoplasm

9.2

Skeletal muscle fibers have contractile myofibrils containing hundreds to thousands of sarcomeres

- skeletal muscle fiber
- myofibril
- myofilaments
- thin filaments
- thick filaments
- sarcomeres
- H band
- M line
- zone of overlap
- A band
- I band
- Z lines
- actinins
- transmembrane potential
- transverse tubules (T tubules)
- sarcoplasmic reticulum (SR)
- terminal cisternae
- triad

● = *Term boldfaced in this module*

9.3

The sliding filament theory of muscle contraction involves thin and thick filaments

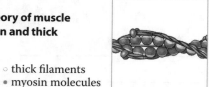

- ○ thin filaments
- • actinin
- • F-actin
- • nebulin
- • G-actin
- • active site
- • tropomyosin
- • troponin

- ○ thick filaments
- • myosin molecules
- • titin
- • tail
- • free head
- • sliding filament theory

9.4

A skeletal muscle fiber contracts when stimulated by a motor neuron

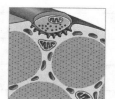

- • neuromuscular junction (NMJ)
- • synaptic terminal
- • motor end plate
- • acetylcholine (ACh)
- • acetylcholinesterase (AChE)

- • junctional folds
- • synaptic cleft
- • action potential
- • excitation– contraction coupling

9.5

A muscle fiber contraction uses ATP in a cycle that is repeated for the duration of the contraction

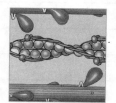

- • contraction cycle
- ○ active-site exposure
- ○ cross-bridge formation
- ○ myosin head pivoting
- ○ cross-bridge detachment
- ○ myosin reactivation

• = *Term boldfaced in this module*

Functional Properties of Skeletal Muscle Tissue

○ neural control
○ excitation–contraction coupling
• tension

_____

_____

9.6

Tension production is greatest when a muscle is stimulated at its optimal length

• optimal resting length
• myogram
• twitch
• latent period
• contraction phase
• relaxation phase

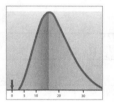

_____

_____

9.7

The peak tension developed by a skeletal muscle depends on the frequency of stimulation and the number of muscle fibers stimulated

• treppe
• wave summation
• incomplete tetanus
• complete tetanus
• motor unit

• recruitment
• asynchronous motor unit summation
• muscle tone

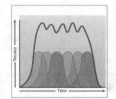

_____

_____

• = _Term boldfaced in this module_

9.8

Muscle contractions may be isotonic or isometric; isotonic contractions may be concentric or eccentric

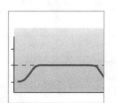

- isotonic contraction
- concentric contraction
- eccentric contraction
- isometric contraction

9.9

Muscle contraction requires large amounts of ATP that may be produced aerobically or anaerobically

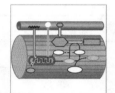

- glycolysis
  - pyruvate
- aerobic metabolism
- creatine phosphate (CP)
- lactic acid
- lactate

9.10

Muscles are subject to fatigue and may require an extended recovery period

- fatigued
- recovery period
- Cori cycle
- oxygen debt
- excess postexercise oxygen consumption (EPOC)

• = *Term boldfaced in this module*

9.11

Fast, slow, and intermediate skeletal muscle fibers differ in size, internal structure, metabolism, and resistance to fatigue

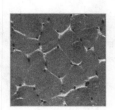

- fast fibers
- slow fibers
- myoglobin
- intermediate fibers

_____

_____

_____

_____

_____

_____

_____

_____

9.12

Many factors can result in muscle hypertrophy, atrophy, or paralysis

- hypertrophy
- atrophy
- polio
- paralysis
- tetanus
- botulism
- myasthenia gravis
- rigor mortis

_____

_____

_____

_____

_____

_____

_____

_____

_____

• = *Term boldfaced in this module*

1. Short answer

Label the pennate muscles in the following diagram, and for each indicate the type of pennate muscle based on the relationship between the tendon(s) and fascicle organization.

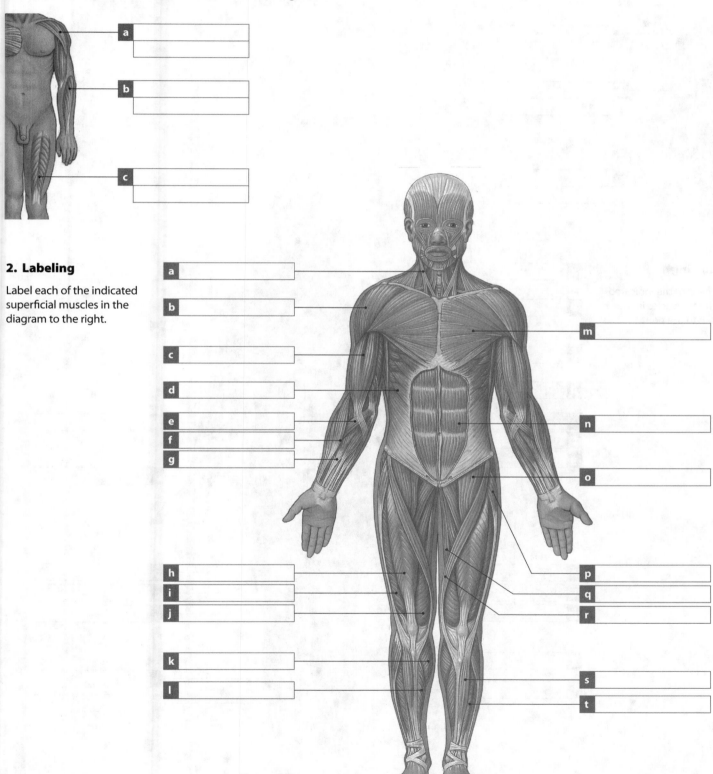

a

b

c

2. Labeling

Label each of the indicated superficial muscles in the diagram to the right.

a

b

c

d

e

f

g

h

i

j

k

l

m

n

o

p

q

r

s

t

1. Short answer

Label the pennate muscles in the following diagram, and for each indicate the type of pennate muscle based on the relationship between the tendon(s) and fascicle organization.

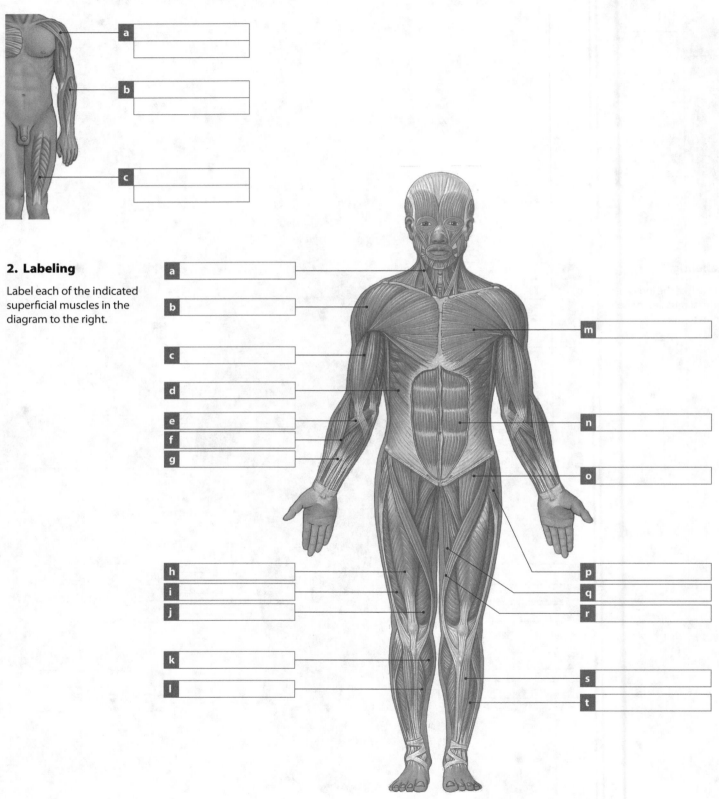

a

b

c

2. Labeling

Label each of the indicated superficial muscles in the diagram to the right.

a

b

c

d

e

f

g

h

i

j

k

l

m

n

o

p

q

r

s

t

1. Labeling

Label each of the indicated muscles of the face in the following diagram.

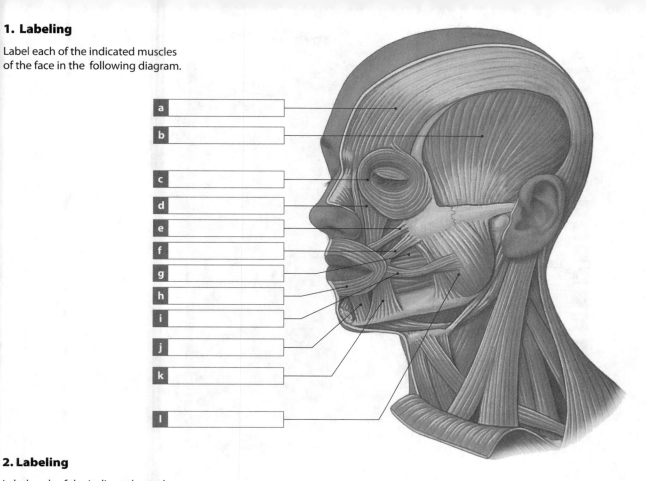

a
b
c
d
e
f
g
h
i
j
k
l

2. Labeling

Label each of the indicated muscles of the neck in the following diagram.

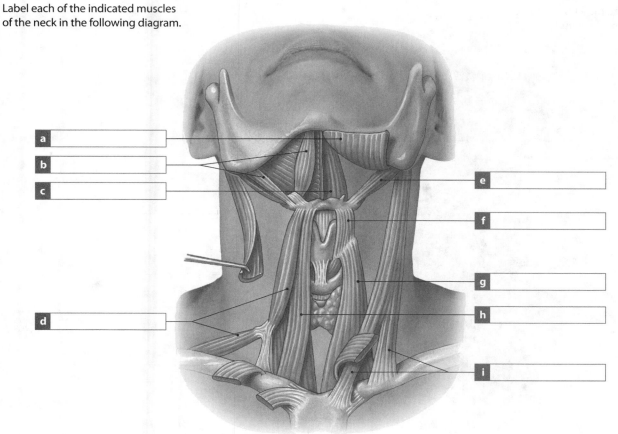

a
b
c
d
e
f
g
h
i

1. Labeling

Label each of the indicated muscles of the face in the following diagram.

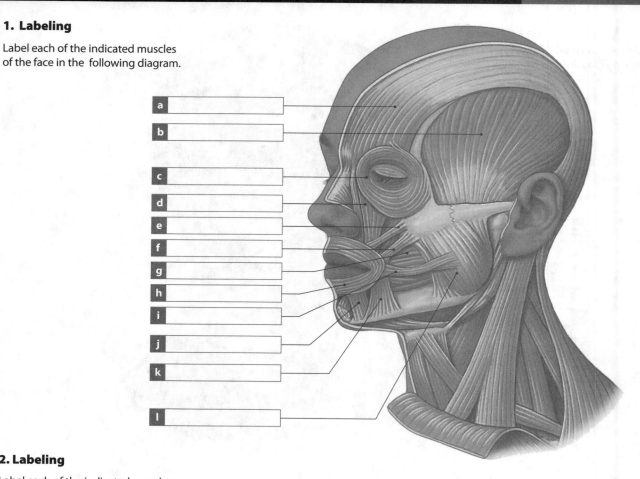

a
b
c
d
e
f
g
h
i
j
k
l

2. Labeling

Label each of the indicated muscles of the neck in the following diagram.

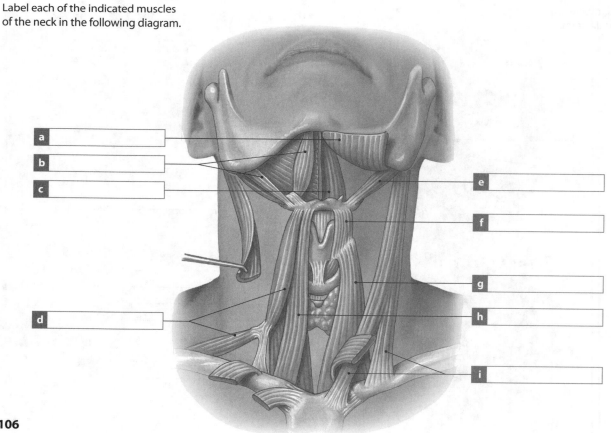

a
b
c
d
e
f
g
h
i

1. Labeling

Label each of the indicated muscles that move the forearm and hand in the diagram at right.

a [ ]

b [ ]

c [ ]

d [ ]

e [ ]

f [ ]

2. Labeling

Label each of the indicated muscles that move the thigh and leg in the diagram below.

a [ ]

b [ ]

c [ ]

d [ ]

e [ ]

f [ ]

g [ ]

h [ ]

i [ ]

j [ ]

h [ ]

3. Labeling

Label each of the indicated muscles that move the foot and toe in the diagram below.

a [ ]

b [ ]

c [ ]

d [ ]

e [ ]

f [ ]

g [ ]

i [ ]

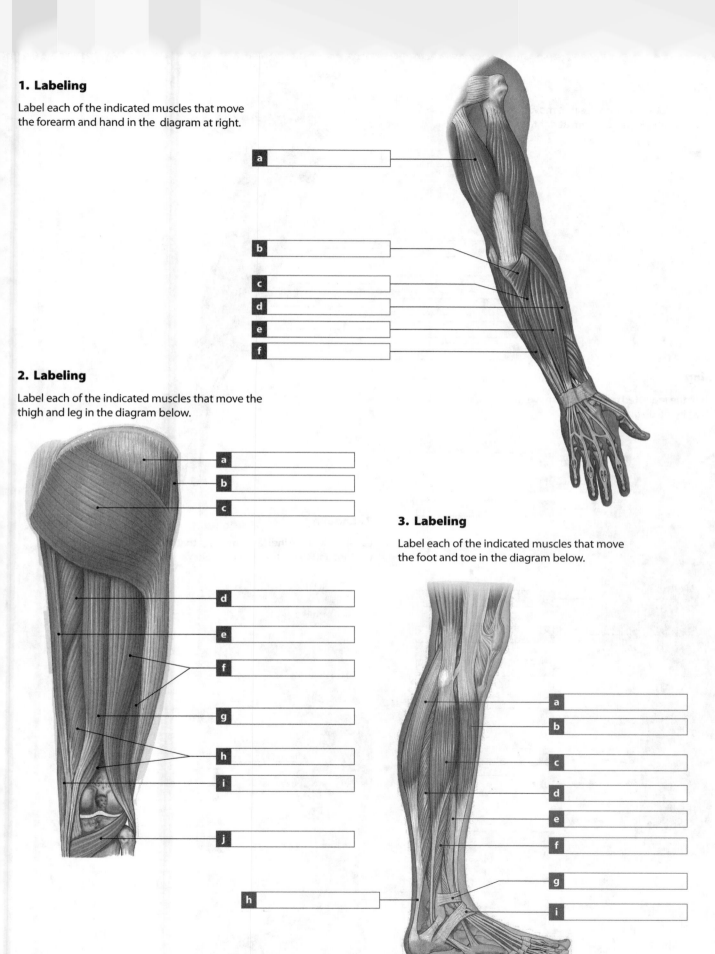

107

1. Labeling

Label each of the indicated muscles that move the forearm and hand in the diagram at right.

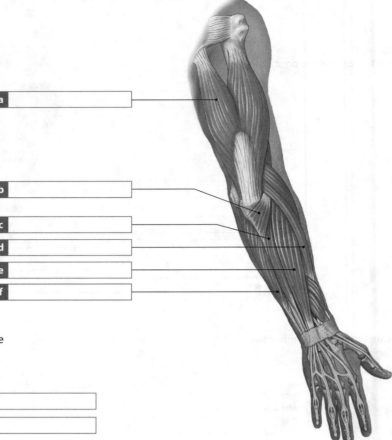

a

b

c

d

e

f

2. Labeling

Label each of the indicated muscles that move the thigh and leg in the diagram below.

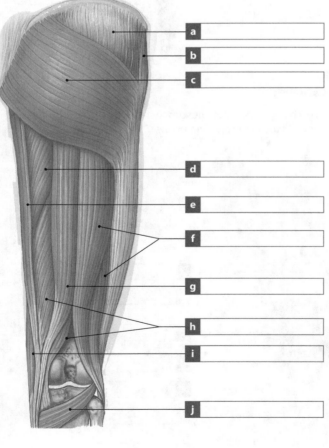

a

b

c

d

e

f

g

h

i

j

3. Labeling

Label each of the indicated muscles that move the foot and toe in the diagram below.

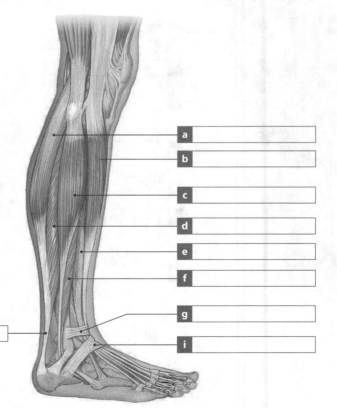

a

b

c

d

e

f

g

i

Visual Outline with Key Terms

Summarize the content of each module using the terms in the order provided.

SECTION 1

**Functional Organization
of the Muscular System**

- muscular system
- axial muscles
- appendicular muscles

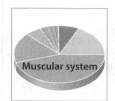

_____

_____

_____

_____

_____

_____

_____

_____

10.1

**Muscular power and range of motion are
influenced by fascicle organization and
leverage**

- parallel muscle
- body
- convergent muscle
- pennate muscle
- unipennate
- bipennate
- multipennate
- circular muscle
  (sphincter)
- first-class lever
- second-class lever
- third-class lever

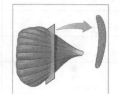

_____

_____

_____

_____

_____

_____

_____

_____

_____

_____

10.2

**The names of muscles can provide clues
to their appearance and/or function**

- origin
- intermuscular septa
- insertion
- action
- agonist
- synergist
- antagonist

_____

_____

_____

_____

_____

_____

_____

• = _Term boldfaced in this module_

10.3

The skeletal muscles can be assigned to the axial division or the appendicular division based on origins and functions

- axial muscles
- appendicular muscles

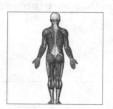

_____

_____

_____

_____

_____

_____

_____

SECTION 2

The Axial Muscles

- ○ muscles of the head and neck
- ○ muscles of the vertebral column
- ○ muscles of the trunk
- ○ muscles of the pelvic floor

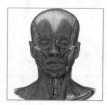

_____

_____

_____

_____

_____

10.4

The muscles of facial expression are important in eating and useful for communication

- muscles of facial expression
- epicranial aponeurosis
- buccinator
- depressor labii inferioris
- levator labii superioris
- levator anguli oris
- mentalis
- orbicularis oris
- risorius
- depressor anguli oris
- zygomaticus major
- zygomaticus minor
- corrugator supercilii
- levator palpebrae superioris
- orbicularis oculi
- procerus
- nasalis
- occipitofrontalis
- platysma

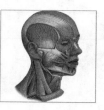

_____

_____

_____

_____

_____

_____

_____

_____

_____

• = _Term boldfaced in this module_

10.5

The extrinsic eye muscles position the eye, and the muscles of mastication move the lower jaw

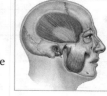

- inferior rectus
- medial rectus
- superior rectus
- lateral rectus
- inferior oblique
- superior oblique
- masseter
- temporalis
- pterygoids

10.6

The muscles of the tongue are closely associated with the muscles of the pharynx and neck

- genioglossus
- hyoglossus
- palatoglossus
- styloglossus
- pharyngeal constrictors
- laryngeal elevators
- palatal muscles
- digastric
- geniohyoid
- mylohyoid
- omohyoid
- sternohyoid
- sternothyroid
- stylohyoid
- thyrohyoid
- sternocleidomastoid

10.7

The muscles of the vertebral column support and align the axial skeleton

- muscles of the vertebral column
- spinal extensors
- splenius
- erector spinae muscles
- spinalis group
- longissimus group
- iliocostalis group
- semispinalis group
- spinal flexors

• = *Term boldfaced in this module*

10.8

The oblique and rectus muscles form the muscular walls of the trunk

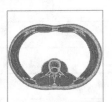

- oblique group
- scalenes
- external intercostals
- internal intercostals
- transversus thoracis
- external oblique
- internal oblique
- transversus abdominis
- rectus group
- diaphragm
- rectus abdominis

10.9

The muscles of the pelvic floor support the organs of the abdominopelvic cavity

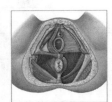

- perineum
- urogenital triangle
- bulbospongiosus
- ischiocavernosus
- superficial transverse perineal
- urogenital diaphragm
- deep transverse perineal
- external urethral sphincter
- anal triangle
- pelvic diaphragm
- coccygeus
- levator ani
- iliococcygeus
- pubococcygeus
- external anal sphincter

SECTION 3

The Appendicular Muscles

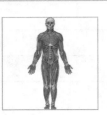

- upper limb
- lower limb

● = *Term boldfaced in this module*

10.10

The largest appendicular muscles originate on the trunk

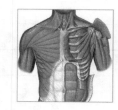

- deltoid
- pectoralis major
- latissimus dorsi
- serratus anterior
- tensor fasciae latae
- sartorius
- rectus femoris
- trapezius
- subclavius
- pectoralis minor
- subscapularis

- coracobrachialis
- biceps brachii
- teres major
- gluteus medius
- iliopsoas
- pectineus
- adductor longus
- gracilis
- infraspinatus
- teres minor
- triceps brachii

- latissimus dorsi
- gluteus maximus
- levator scapulae
- supraspinatus
- rhomboid minor
- rhomboid major

10.11

Muscles that position the pectoral girdle originate on the occipital bone, superior vertebrae, and ribs

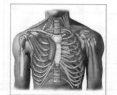

- pectoral girdle
- levator scapulae
- pectoralis minor
- rhomboid major

- rhomboid minor
- serratus anterior
- subclavius
- trapezius

10.12

Muscles that move the arm originate on the clavicle, scapula, thoracic cage, and vertebral column

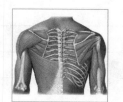

- deltoid
- supraspinatus
- subscapularis
- teres major
- infraspinatus
- teres minor

- coracobrachialis
- pectoralis major
- latissimus dorsi
- action line
- rotator cuff

• = *Term boldfaced in this module*

**Muscles that move the forearm and
hand originate on the scapula, humerus,
radius, or ulna**

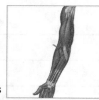

- extensor retinaculum
- flexor retinaculum
- biceps brachii
- brachialis
- brachioradialis
- anconeus
- triceps brachii
- pronator quadratus
- pronator teres
- supinator
- flexor carpi radialis
- flexor carpi ulnaris
- palmaris longus
- extensor carpi radialis longus
- extensor carpi radialis brevis
- extensor carpi ulnaris
- synovial tendon sheaths
- carpal tunnel syndrome

**Muscles that move the hand and fingers
originate on the humerus, radius, ulna,
and interosseus membrane**

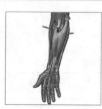

- abductor pollicis longus
- extensor digitorum
- extensor pollicis brevis
- extensor pollicis longus
- extensor indicis
- extensor digiti minimi
- flexor digitorum superficialis
- flexor digitorum profundus
- flexor pollicis longus

**The intrinsic muscles of the hand
originate on the carpal and metacarpal
bones and associated tendons and
ligaments**

- ○ intrinsic muscles of the hand
- ○ intrinsic muscles of the thumb
- palmaris brevis
- adductor pollicis
- palmar interosseus
- abductor pollicis brevis
- dorsal interosseus
- abductor digiti minimi
- flexor pollicis brevis
- lumbricals
- flexor digiti minimi brevis
- opponens pollicis
- opponens digiti minimi

• = *Term boldfaced in this module*

10.16

The muscles that move the thigh
originate on the pelvis and associated
ligaments and fasciae

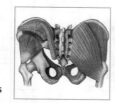

- gluteal group
- gluteus maximus
- gluteus medius
- gluteus minimus
- tensor fasciae latae
- lateral rotator group
- obturator externus
- obturator internus
- piriformis
- superior gemellus
- inferior gemellus
- quadratus femoris
- adductor group
- adductor brevis
- adductor longus
- adductor magnus
- pectineus
- gracilis
- iliopsoas group
- iliacus
- psoas major

10.17

The muscles that move the leg originate
on the pelvis and femur

- flexors of the knee
- biceps femoris
- semimembranosus
- semitendinosus
- sartorius
- popliteus
- extensors of the
  knee
- quadriceps muscles
  (quadriceps femoris)
- rectus femoris
- vastus intermedius
- vastus lateralis
- vastus medialis

10.18

The extrinsic muscles that move the foot
and toes originate on the tibia and fibula

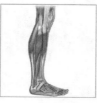

- tibialis anterior
- gastrocnemius
- fibularis brevis
- fibularis longus
- plantaris
- soleus
- tibialis posterior
- flexor digitorum
  longus
- flexor hallucis longus
- extensor digitorum
  longus
- extensor hallucis
  longus
- superior
  retinaculum
- inferior retinaculum

• = *Term boldfaced in this module*

The intrinsic muscles of the foot originate on the tarsal and metatarsal bones and associated tendons and ligaments

- intrinsic muscles of the foot
- flexor hallucis brevis
- flexor digitorum brevis
- quadratus plantae
- lumbrical (4)
- flexor digiti minimi brevis
- extensor digitorum brevis
- extensor hallucis brevis
- adductor hallucis
- abductor hallucis
- plantar interosseus (3)
- dorsal interosseus (4)
- abductor digiti minimi

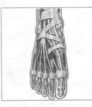

The deep fascia separates the limb muscles into separate compartments

- compartments
- compartment syndrome

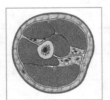

• = *Term boldfaced in this module*

1. Vocabulary

In the space provided, write the boldfaced terms introduced in this section that contain the indicated word part.

Word Part	Meaning	Term(s)
a neur-	nerve	_____
b dendr-	tree	_____
c ef-	away from	_____
d af-	toward	_____

2. Labeling

Label each of the structures in the following diagram of a neuron.

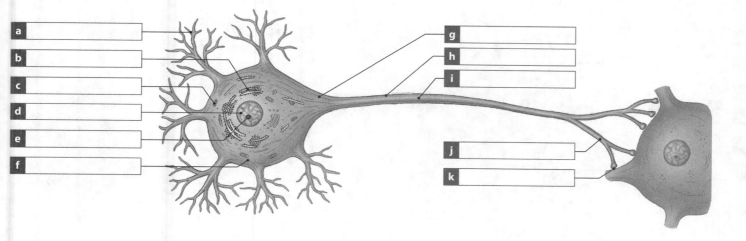

3. Labeling

Label the types of neurons depicted below.

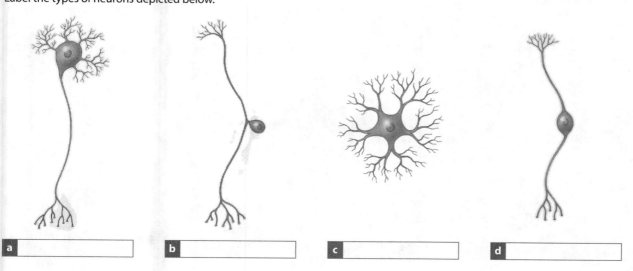

1. Vocabulary

In the space provided, write the boldfaced terms introduced in this section that contain the indicated word part.

Word Part	Meaning	Term(s)
a neur-	nerve	_____
b dendr-	tree	_____
c ef-	away from	_____
d af-	toward	_____

2. Labeling

Label each of the structures in the following diagram of a neuron.

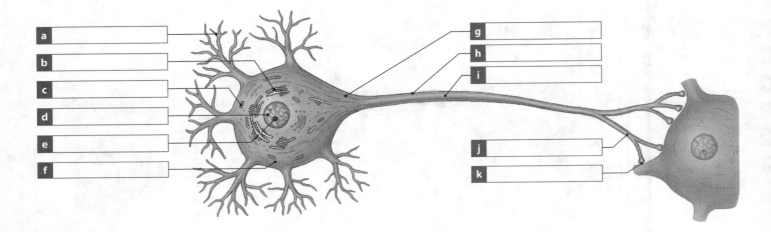

a _____
b _____
c _____
d _____
e _____
f _____
g _____
h _____
i _____
j _____
k _____

3. Labeling

Label the types of neurons depicted below.

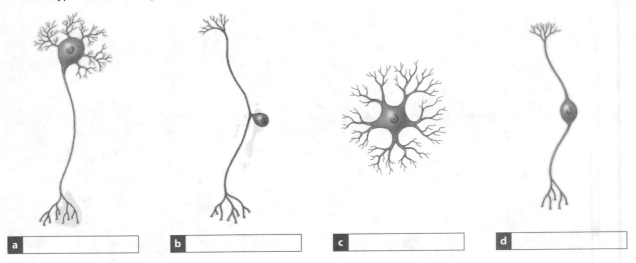

a _____
b _____
c _____
d _____

1. Vocabulary

Write the boldfaced term introduced in this section in the blank next to the term's definition.

a _____ A propagated change in the transmembrane potential

b _____ A synapse in which the presynaptic and postsynaptic neuronal membranes are locked together by gap junctions

c _____ The transmembrane potential of a nonstimulated cell

d _____ Ion channels that open or close in response to specific stimuli

e _____ Chemical synapses that release acetylcholine

f _____ A shift in the transmembrane potential from −70 mV to −85 mV

g _____ The movement of positive charges parallel to the inner and outer membrane surfaces

h _____ A shift in the transmembrane potential from −70 mV to +30 mV

2. Short answer

For the following diagram of a cholinergic synapse, write the names of components a–f in the boxes at left, and then fill in the table at right with descriptions of the events represented by g–l.

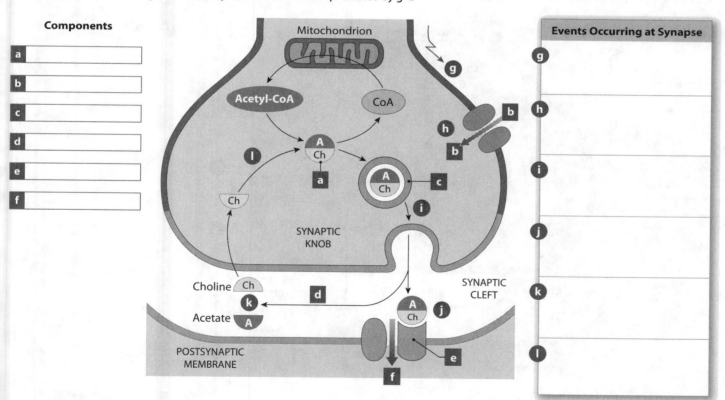

Components

a
b
c
d
e
f

Events Occurring at Synapse
g
h
i
j
k
l

3. Section integration

Guillain-Barré (*ghee-yan bah-ray*) syndrome is a degeneration of myelin sheaths that ultimately may result in paralysis. Propose a mechanism by which myelin sheath degeneration can cause muscular paralysis.

_____

_____

1. Vocabulary

Write the boldfaced term introduced in this section in the blank next to the term's definition.

a	_____	A propagated change in the transmembrane potential
b	_____	A synapse in which the presynaptic and postsynaptic neuronal membranes are locked together by gap junctions
c	_____	The transmembrane potential of a nonstimulated cell
d	_____	Ion channels that open or close in response to specific stimuli
e	_____	Chemical synapses that release acetylcholine
f	_____	A shift in the transmembrane potential from −70 mV to −85 mV
g	_____	The movement of positive charges parallel to the inner and outer membrane surfaces
h	_____	A shift in the transmembrane potential from −70 mV to +30 mV

2. Short answer

For the following diagram of a cholinergic synapse, write the names of components a–f in the boxes at left, and then fill in the table at right with descriptions of the events represented by g–l.

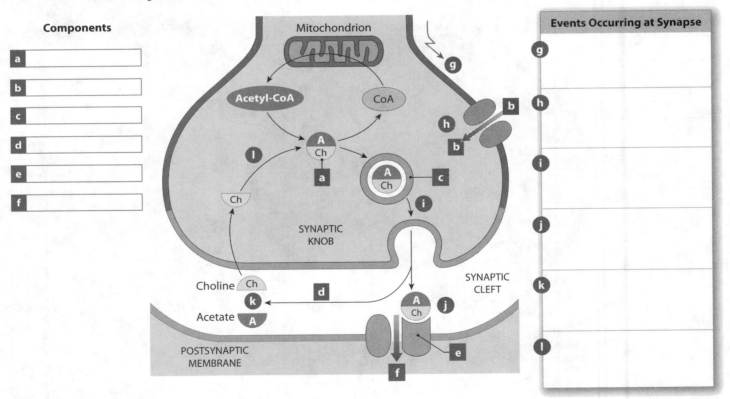

3. Section integration

Guillain-Barré *(ghee-yan bah-ray)* syndrome is a degeneration of myelin sheaths that ultimately may result in paralysis. Propose a mechanism by which myelin sheath degeneration can cause muscular paralysis.

_____

_____

Visual Outline with Key Terms

Summarize the content of each module using the terms in the order provided.

Neurons and Neuroglia

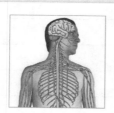

- receptors
- visceral sensory receptors
- somatic sensory receptors
- special sensory receptors
- sensory division
- peripheral nervous system (PNS)
- information processing
- central nervous system (CNS)
- motor division
- somatic nervous system (SNS)
- autonomic nervous system (ANS)
- effectors

_____
_____
_____
_____
_____
_____
_____
_____
_____
_____

11.1

Neurons are nerve cells specialized for intercellular communication

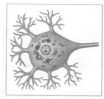

- dendrites
- cell body
- axons
- neurotubules
- axoplasmic transport
- retrograde flow
- dendritic spines
- axon hillock
- axolemma
- axoplasm
- nissl bodies
- perikaryon
- neurofilaments
- neurofibrils
- telodendria/ telodendrion
- synaptic terminals
- synapse
- presynaptic cell
- postsynaptic cell
- neurotransmitters
- synaptic cleft
- synaptic knob
- synaptic vesicles
- presynaptic membrane
- postsynaptic membrane
- collateral branches

_____
_____
_____
_____
_____
_____
_____
_____
_____
_____
_____
_____
_____

• = _Term boldfaced in this module_

11.2

Neurons may be classified on the basis of structure or function

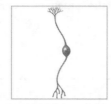

- anaxonic neurons
- bipolar neurons
- unipolar neuron
- initial segment
- multipolar neurons
- sensory neurons
- interneurons
- motor neurons
- sensory receptors
- interoceptors
- proprioceptors
- exteroceptors
- afferent fibers
- sensory ganglia
- visceral sensory neurons
- somatic sensory neurons
- somatic motor neurons
- peripheral nerve
- visceral motor neurons
- efferent fibers
- autonomic ganglia
- somatic effectors
- visceral effectors

11.3

Oligodendrocytes, astrocytes, ependymal cells, and microglia are neuroglia of the CNS

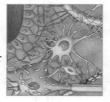

- neuroglia (glial cells)
- ependymal cells
- ependyma
- cerebrospinal fluid (CSF)
- microglia
- astrocytes
- blood–brain barrier
- oligodendrocytes
- myelin
- myelin sheath
- myelinated
- internodes
- nodes
- white matter
- unmyelinated axons
- gray matter

11.4

Schwann cells and satellite cells are the neuroglia of the PNS

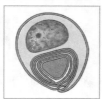

- Schwann cells
- neurilemma
- satellite cells
- Wallerian degeneration

Neurophysiology

- transmembrane potential
- resting potential
- graded potential
- action potential
- synaptic activity
- information processing

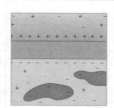

11.5

The resting potential is the transmembrane potential of an undisturbed cell

- leak channels
- resting potential
- volt (V)
- millivolt (mV)

- chemical gradient
- electrical gradient
- equilibrium potential

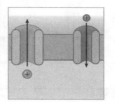

11.6

Gated channels can change the permeability of the plasma membrane

- gated channels
- chemically gated channels
- voltage-gated channels
- mechanically gated channels

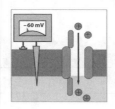

11.7

Graded potentials are localized changes in the transmembrane potential

- graded potentials
- depolarization
- local current
- repolarization
- hyperpolarization

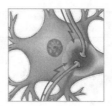

• = Term boldfaced in this module

11.8

An action potential begins with the opening of voltage-gated sodium ion channels

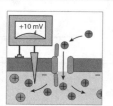

- action potentials
- threshold
- sodium channel inactivation
- absolute refractory period
- relative refractory period
- all-or-none principle

11.9

Action potentials may affect adjacent portions of the plasma membrane through continuous propagation or saltatory propagation

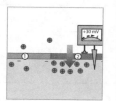

- propagation
- continuous propagation
- saltatory propagation

11.10

At a synapse, information travels from the presynaptic cell to the postsynaptic cell

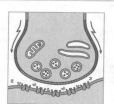

- synapse
- chemical synapses
- cholinergic synapses
- synaptic fatigue
- synaptic delay
- electrical synapse

• = *Term boldfaced in this module*

11.11

Postsynaptic potentials are responsible for information processing in a neuron

- postsynaptic potentials
- excitatory postsynaptic potential (EPSP)
- facilitated
- inhibitory postsynaptic potential (IPSP)
- summation
- temporal summation
- spatial summation

11.12

Information processing involves interacting groups of neurons, and information is encoded in the frequency and pattern of action potentials

- regulatory neurons
- G proteins
- second messengers

• = *Term boldfaced in this module*

1. Labeling

Label each of the structures in the following cross-sectional diagram of the spinal cord.

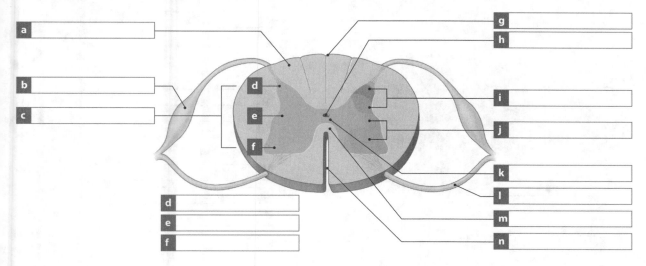

a
b
c
d
e
f

g
h
i
j
k
l
m
n

2. Labeling

Label the views (a, e) and the nerves that innervate the indicated regions of the hands (b–d).

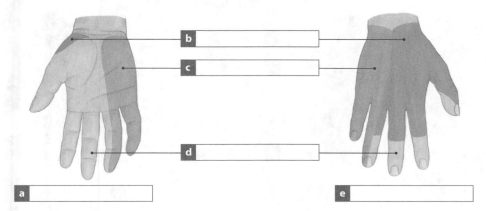

a
b
c
d
e

3. Vocabulary

Write the boldfaced term introduced in this section that matches each of the following definitions.

a _____ The regions of white matter in the spinal cord

b _____ The terminal portion of the spinal cord

c _____ Bundles of axons in the PNS plus their associated blood vessels and connective tissues

d _____ Specialized membranes that provide stability and support for the spinal cord and brain

e _____ The complex made up of the filum terminale and the long dorsal and ventral spinal nerve roots inferior to the spinal cord

f _____ The plexus from which the radial, median, and ulnar nerves originate

g _____ The outermost covering of the spinal cord

h _____ The connective tissue partition that separates adjacent bundles of nerve fibers in a spinal nerve or a peripheral nerve

i _____ A bundle of unmyelinated, postganglionic fibers that innervates glands and smooth muscles in the body wall or limbs

1. Labeling

Label each of the structures in the following cross-sectional diagram of the spinal cord.

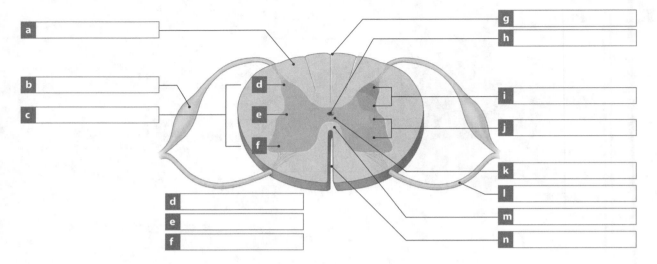

2. Labeling

Label the views (a, e) and the nerves that innervate the indicated regions of the hands (b–d).

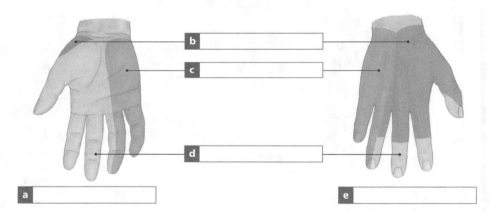

3. Vocabulary

Write the boldfaced term introduced in this section that matches each of the following definitions.

a _____ The regions of white matter in the spinal cord

b _____ The terminal portion of the spinal cord

c _____ Bundles of axons in the PNS plus their associated blood vessels and connective tissues

d _____ Specialized membranes that provide stability and support for the spinal cord and brain

e _____ The complex made up of the filum terminale and the long dorsal and ventral spinal nerve roots inferior to the spinal cord

f _____ The plexus from which the radial, median, and ulnar nerves originate

g _____ The outermost covering of the spinal cord

h _____ The connective tissue partition that separates adjacent bundles of nerve fibers in a spinal nerve or a peripheral nerve

i _____ A bundle of unmyelinated, postganglionic fibers that innervates glands and smooth muscles in the body wall or limbs

1. Labeling

Label the neural circuit patterns in the following diagrams.

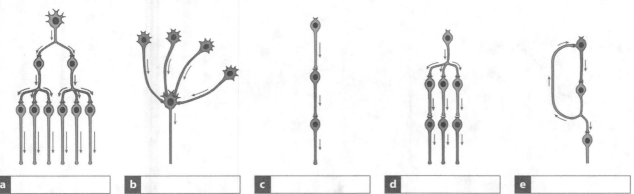

a		b		c		d		e	

2. Labeling

Label the indicated structures in the accompanying diagram of a reflex arc.

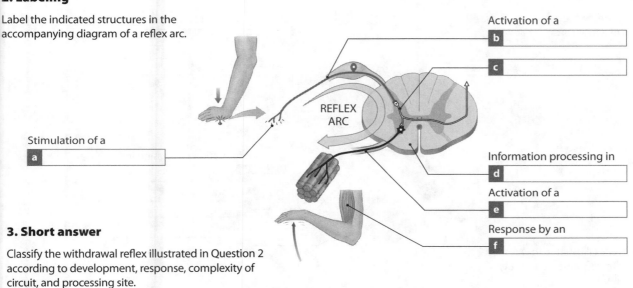

REFLEX ARC

Activation of a
b

c

Stimulation of a
a

Information processing in
d

Activation of a
e

Response by an
f

3. Short answer

Classify the withdrawal reflex illustrated in Question 2 according to development, response, complexity of circuit, and processing site.

_____

_____

4. Vocabulary

Write the boldfaced term introduced in this section that matches each of the following definitions.

a _____ A reflex in which the sensory stimulus and motor response occur on the same side of the body

b _____ Reflexes that move affected parts of the body away from a stimulus

c _____ Controls the sensitivity of a muscle spindle

d _____ Withdrawal reflex that affects the muscles of a limb

e _____ Autonomic reflexes that adjust nonskeletal muscles, glands, and adipose tissue

f _____ Type of reflexes enhanced by repetition

g _____ A reflex in which the motor response occurs on the side opposite to the stimulus

h _____ Process involved in preventing opposing muscles from contracting during a reflex

i _____ The enhancement of reflexes due to facilitation of motor neurons

1. Labeling

Label the neural circuit patterns in the following diagrams.

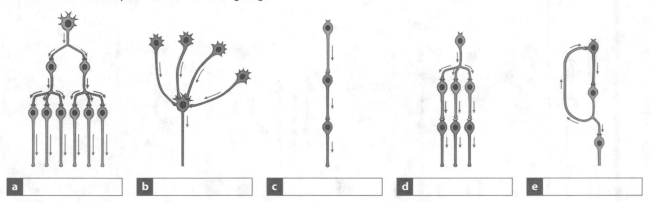

a _____ b _____ c _____ d _____ e _____

2. Labeling

Label the indicated structures in the accompanying diagram of a reflex arc.

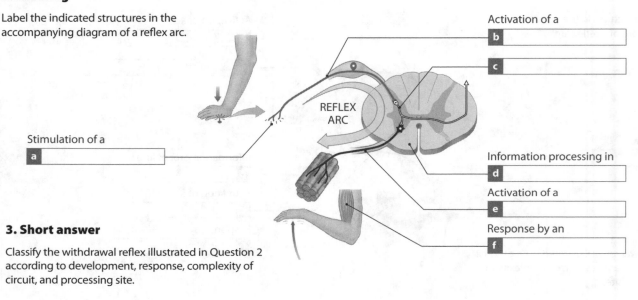

REFLEX ARC

Stimulation of a
a _____

Activation of a
b _____

c _____

Information processing in
d _____

Activation of a
e _____

Response by an
f _____

3. Short answer

Classify the withdrawal reflex illustrated in Question 2 according to development, response, complexity of circuit, and processing site.

_____

_____

4. Vocabulary

Write the boldfaced term introduced in this section that matches each of the following definitions.

a _____ A reflex in which the sensory stimulus and motor response occur on the same side of the body

b _____ Reflexes that move affected parts of the body away from a stimulus

c _____ Controls the sensitivity of a muscle spindle

d _____ Withdrawal reflex that affects the muscles of a limb

e _____ Autonomic reflexes that adjust nonskeletal muscles, glands, and adipose tissue

f _____ Type of reflexes enhanced by repetition

g _____ A reflex in which the motor response occurs on the side opposite to the stimulus

h _____ Process involved in preventing opposing muscles from contracting during a reflex

i _____ The enhancement of reflexes due to facilitation of motor neurons

Visual Outline with Key Terms

Summarize the content of each module using the terms in the order provided.

SECTION 1

The Functional Organization of the Spinal Cord

- spinal cord

_____

_____

_____

_____

_____

_____

_____

12.1

The spinal cord contains gray matter and white matter

- cervical enlargement
- lumbar enlargement
- conus medullaris
- cauda equina
- filum terminale
- posterior median sulcus
- dorsal root
- dorsal root ganglion
- spinal nerve
- ventral root
- anterior median fissure

- white matter
- gray matter

_____

_____

_____

_____

_____

_____

_____

12.2

The spinal cord is surrounded by the meninges, which consist of the dura mater, arachnoid mater, and pia mater

- spinal meninges
- pia mater
- arachnoid mater
- subarachnoid space
- dura mater
- cerebrospinal fluid (CSF)
- epidural space
- denticulate ligaments
- lumbar puncture

_____

_____

_____

_____

_____

_____

_____

• = _Term boldfaced in this module_

12.3

Gray matter is the region of integration, and white matter carries information

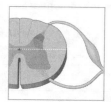

- horns
- posterior gray horn
- lateral gray horn
- anterior gray horn
- gray commissures
- nuclei
- sensory nuclei
- motor nuclei
- columns
- tracts
- ascending tracts
- descending tracts
- posterior white column
- lateral white column
- anterior white column
- anterior white commissure

12.4

Spinal nerves have a relatively consistent anatomical structure and pattern of distribution

- epineurium
- perineurium
- fascicles
- endoneurium
- rami
- sympathetic division
- dorsal ramus
- ventral ramus
- communicating rami
- dermatome
- shingles

12.5

Each ramus (branch) of a spinal nerve provides motor and sensory innervation to a specific region

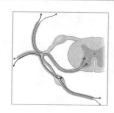

- dorsal ramus
- ventral ramus
- white ramus
- gray ramus
- rami communicantes
- sympathetic nerve

● = *Term boldfaced in this module*

12.6

Spinal nerves form nerve plexuses that innervate the skin and skeletal muscles; the cervical plexus is the smallest of these nerve plexuses

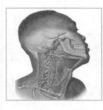

- nerve plexus
- cervical plexus
- brachial plexus
- lumbar plexus
- sacral plexus
- phrenic nerve
- ansa cervicalis

- lesser occipital nerve
- transverse cervical nerve
- supraclavicular nerve

- great auricular nerve
- cervical nerves

12.7

The brachial plexus innervates the pectoral girdle and upper limbs

- trunks
- cords
- dorsal scapular nerve
- long thoracic nerve
- suprascapular nerve
- pectoral nerves
- subscapular nerves
- thoracodorsal nerve

- axillary nerve
- medial antebrachial cutaneous nerve
- radial nerve
- musculocutaneous nerve

- median nerve
- ulnar nerve

12.8

The lumbar and sacral plexuses innervate the skin and skeletal muscles of the trunk and lower limbs

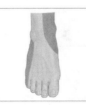

- lumbar plexus
- sacral plexus
- sciatic nerve
- iliohypogastric nerve
- ilioinguinal nerve
- genitofemoral nerve
- lateral femoral cutaneous nerve

- femoral nerve
- obturator nerve
- saphenous nerve
- gluteal nerves
- posterior femoral cutaneous nerve
- tibial nerve

- fibular nerve
- pudendal nerve
- sural nerve

● = *Term boldfaced in this module*

SECTION 2

An Introduction to Reflexes

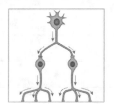

- neuronal pools
- neural circuit
- divergence
- parallel processing
- serial processing
- convergence
- reverberation
- reflexes

_____

_____

12.9

Reflexes are rapid, automatic responses to stimuli

- reflexes
- reflex arc
- innate reflexes
- acquired reflexes
- somatic reflexes
- visceral reflexes
- polysynaptic reflexes
- monosynaptic reflexes
- spinal reflexes
- intersegmental reflexes
- cranial reflexes

_____

_____

12.10

The stretch reflex is a monosynaptic reflex involving muscle spindles

- stretch reflex
- muscle spindles
- intrafusal muscle fibers
- gamma motor neurons
- postural reflexes

_____

_____

• = *Term boldfaced in this module*

12.11

Withdrawal reflexes and crossed extensor reflexes are polysynaptic reflexes

- withdrawal reflexes
- flexor reflex
- reciprocal inhibition
- ipsilateral reflex arcs
- crossed extensor reflex
- contralateral reflex arc
- polysynaptic reflexes

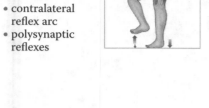
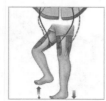

12.12

The brain can inhibit or facilitate spinal reflexes, and reflexes can be used to determine the location and severity of damage to the CNS

- reinforcement
- biceps reflex
- triceps reflex
- ankle-jerk reflex
- Babinski sign
- plantar reflex
- abdominal reflex

• = *Term boldfaced in this module*

1. Labeling

Label the structures in the accompanying figure of a lateral view of the human brain.

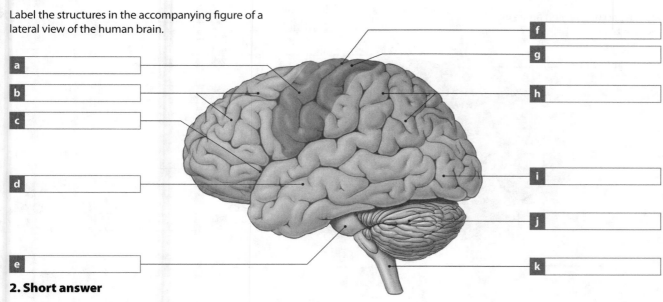

a _____

b _____

c _____

d _____

e _____

f _____

g _____

h _____

i _____

j _____

k _____

2. Short answer

Identify the cranial nerves in the accompanying figure, and indicate the function of each: M = motor, S = sensory, or B = both motor and sensory.

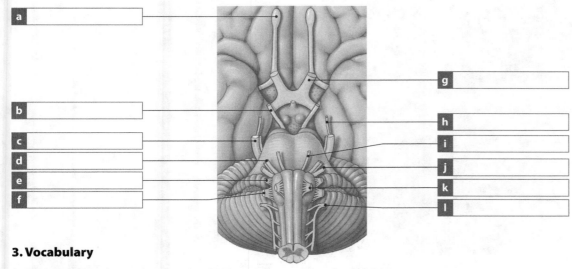

a _____

b _____

c _____

d _____

e _____

f _____

g _____

h _____

i _____

j _____

k _____

l _____

3. Vocabulary

Write the boldfaced term introduced in this section that matches the definition.

a _____ Forms the walls of the diencephalon

b _____ The shortest association fibers in the CNS white matter

c _____ The tract of white matter that connects the hippocampus with the hypothalamus

d _____ The fibers that permit communication between the two cerebral hemispheres

e _____ The nuclei made up of the caudate nucleus and the lentiform nucleus

4. Section integration

Smelling salts may restore consciousness after a person has fainted. The active ingredient of smelling salts is ammonia, and it acts by irritating the lining of the nasal cavity. Propose a mechanism by which smelling salts would raise a person from the unconscious state to the conscious state.

_____

_____

1. Labeling

Label the structures in the accompanying figure of a lateral view of the human brain.

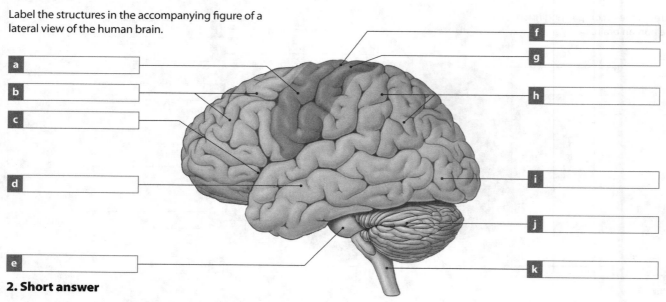

a
b
c
d
e
f
g
h
i
j
k

2. Short answer

Identify the cranial nerves in the accompanying figure, and indicate the function of each: M = motor, S = sensory, or B = both motor and sensory.

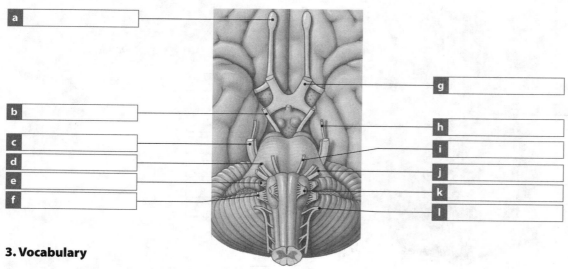

a
b
c
d
e
f
g
h
i
j
k
l

3. Vocabulary

Write the boldfaced term introduced in this section that matches the definition.

a _____ Forms the walls of the diencephalon

b _____ The shortest association fibers in the CNS white matter

c _____ The tract of white matter that connects the hippocampus with the hypothalamus

d _____ The fibers that permit communication between the two cerebral hemispheres

e _____ The nuclei made up of the caudate nucleus and the lentiform nucleus

4. Section integration

Smelling salts may restore consciousness after a person has fainted. The active ingredient of smelling salts is ammonia, and it acts by irritating the lining of the nasal cavity. Propose a mechanism by which smelling salts would raise a person from the unconscious state to the conscious state.

_____

1. Labeling

Label each type of tactile receptor found in the skin.

 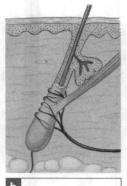 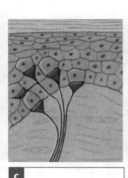 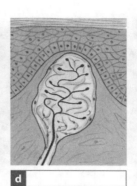  

a _____ b _____ c _____ d _____ e _____ f _____

2. Short answer

Identify the descending and ascending tracts and pathways in the accompanying sectional diagram of the spinal cord, and then describe the general functions of the tracts of each pathway.

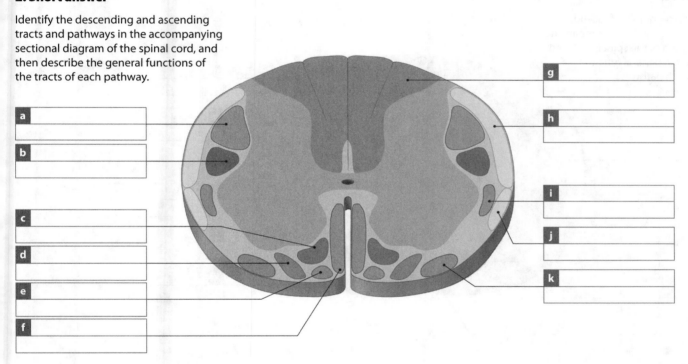

a _____
b _____
c _____
d _____
e _____
f _____

g _____
h _____
i _____
j _____
k _____

3. Short answer

The general organization of the spinal cord is such that motor tracts are **a** _____ (anterior or posterior), and sensory tracts are **b** _____ (anterior or posterior).

4. Section integration

An individual whose primary cortex has been injured retains the ability to walk, maintain balance, and perform other voluntary and involuntary movements. Even though the movements lack precision and are awkward and poorly controlled, why is the ability to walk and maintain balance possible?

_____

_____

1. Labeling

Label each type of tactile receptor found in the skin.

a

b

c

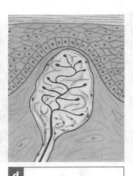

d

e

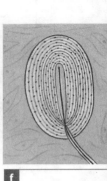

f

2. Short answer

Identify the descending and ascending tracts and pathways in the accompanying sectional diagram of the spinal cord, and then describe the general functions of the tracts of each pathway.

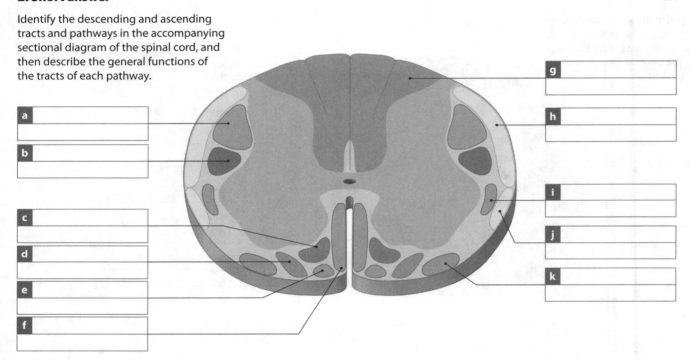

a

b

c

d

e

f

g

h

i

j

k

3. Short answer

The general organization of the spinal cord is such that motor tracts are **a** _____ (anterior or posterior), and sensory tracts are **b** _____ (anterior or posterior).

4. Section integration

An individual whose primary cortex has been injured retains the ability to walk, maintain balance, and perform other voluntary and involuntary movements. Even though the movements lack precision and are awkward and poorly controlled, why is the ability to walk and maintain balance possible?

_____

_____

Visual Outline with Key Terms

Summarize the content of each module using the terms in the order provided.

The Functional Anatomy of the Brain and Cranial Nerves

- neural tube
- neurocoel
- primary brain vesicles
- mesencephalon
- prosencephalon
- rhombencephalon
- secondary brain vesicles
- diencephalon
- telencephalon
- cerebrum
- metencephalon
- cerebellum
- pons

- myelencephalon
- medulla oblongata

_____

_____

_____

_____

_____

_____

_____

_____

_____

13.1

Each region of the brain has distinct structural and functional characteristics

- cerebrum
- cerebral hemispheres
- cerebral cortex
- fissures
- gyri
- sulci
- diencephalon
- thalamus
- hypothalamus
- brain stem
- midbrain
- pons
- medulla oblongata
- cerebellum
- ventricles
- lateral ventricle
- interventricular foramen
- third ventricle

- aqueduct of the midbrain
- fourth ventricle
- corpus callosum
- septum pellucidum

_____

_____

_____

_____

_____

_____

_____

_____

_____

_____

_____

13.2

The brain is protected and supported by the cranial meninges and the cerebrospinal fluid

- cranial meninges
- dura mater
- arachnoid mater
- pia mater
- dural folds
- dural sinuses
- falx cerebri
- superior sagittal sinus
- tentorium cerebelli
- falx cerebelli
- cerebrospinal fluid (CSF)
- choroid plexus
- lateral apertures

- median aperture
- arachnoid granulations

_____

_____

_____

_____

_____

_____

_____

_____

_____

• = _Term boldfaced in this module_

13.3

The medulla oblongata and the pons contain autonomic reflex centers, relay stations, and ascending and descending tracts

- medulla oblongata
- pyramids
- decussation
- autonomic centers
- relay stations
- olive
- nucleus gracilis
- nucleus cuneatus
- olivary nuclei
- solitary nucleus
- ascending and descending tracts
- pons
- tracts
- respiratory centers
- reticular formation
- nuclei associated with cranial nerves (V, VI, VII, VIII)
- apneustic and pneumotaxic centers
- transverse fibers

13.4

The cerebellum coordinates learned and reflexive patterns of muscular activity at the subconscious level

- cerebellum
- anterior lobe
- posterior lobe
- cerebellar cortex
- vermis
- primary fissure
- folia
- Purkinje cells
- cerebellar peduncles
- arbor vitae
- transverse fibers
- ataxia

13.5

The midbrain regulates auditory and visual reflexes and controls alertness

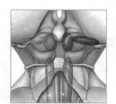

- midbrain
- corpora quadrigemina
- superior colliculus
- inferior colliculus
- reticular activating system (RAS)
- red nucleus
- substantia nigra
- cerebral peduncles
- tectum
- tegmentum

• = *Term boldfaced in this module*

13.6

The diencephalon consists of the epithalamus, thalamus (left and right), and hypothalamus

- epithalamus
- anterior commissure
- optic chiasm
- interthalamic adhesion
- pineal gland
- melatonin
- thalamus
- lateral geniculate nucleus
- optic tract
- medial geniculate nucleus
- projected
- hypothalamus
- preoptic area
- suprachiasmatic nucleus
- infundibulum
- mamillary bodies

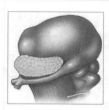

13.7

The limbic system is a functional group of tracts and nuclei located in the cerebrum and diencephalon

- limbic system
- limbic lobe
- cingulate gyrus
- parahippocampal gyrus
- fornix
- amygdaloid body
- hippocampus

13.8

The basal nuclei of the cerebrum perform subconscious adjustment and refinement of ongoing voluntary movements

- basal nuclei
- caudate nucleus
- lentiform nucleus
- globus pallidus
- putamen
- internal capsule

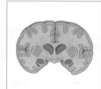

• = *Term boldfaced in this module*

13.9

Superficial landmarks can be used to divide the surface of the cerebral cortex into lobes

- lobes
- precentral gyrus
- central sulcus
- frontal lobe
- parietal lobe
- postcentral gyrus
- lateral sulcus
- temporal lobe
- insula
- parieto-occipital sulcus

13.10

The lobes of the cerebral cortex contain regions with specific functions

- primary motor cortex
- primary sensory cortex
- association area
- motor cortex
- pyramidal cells
- somatic motor association area
- gustatory cortex
- olfactory cortex
- auditory cortex
- primary auditory cortex
- auditory association area
- sensory cortex
- somatic sensory association area
- visual cortex
- primary visual cortex
- visual association area
- integrative centers
- speech center
- prefrontal cortex
- frontal eye field
- general interpretive area
- hemispheric lateralization

13.11

White matter interconnects the cerebral hemispheres, the lobes of each hemisphere, and links the cerebrum to the rest of the brain

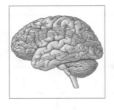

- association fibers
- arcuate fibers
- longitudinal fasciculi
- commissural fibers
- projection fibers
- internal capsule
- corpus callosum
- anterior commissure

• = *Term boldfaced in this module*

13.12

Brain activity can be monitored using external electrodes; the record is called an electroencephalogram, or EEG

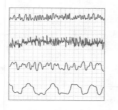

- electroencephalogram (EEG)
- brain waves
- alpha waves
- beta waves
- theta waves
- delta waves
- seizure
- epilepsies

13.13

The twelve pairs of cranial nerves can be classified as sensory, special sensory, motor, or mixed nerves

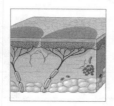

- olfactory nerves (I)
- optic nerves (II)
- oculomotor nerves (III)
- trochlear nerves (IV)
- trigeminal nerves (V)
- abducens nerves (VI)
- facial nerves (VII)
- vestibulocochlear nerves (VIII)
- glossopharyngeal nerves (IX)
- vagus nerves (X)
- accessory nerves (XI)
- hypoglossal nerves (XII)

SECTION 2

Sensory and Motor Pathways

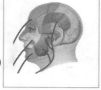

- general senses
- sensation
- perception
- receptive field
- transduction
- labeled line

• = *Term boldfaced in this module*

13.14

Receptors for the general senses can be classified by function and by sensitivity

- nociceptors
- Type A fibers
- fast pain
- Type C fibers
- slow pain
- thermoreceptors
- chemoreceptors
- mechanoreceptors
- proprioceptors
- baroreceptors
- tactile receptors
- fine touch and pressure receptors
- crude touch and pressure receptors
- tonic receptors
- phasic receptors
- adaptation
- peripheral adaptation
- central adaptation

13.15

General sensory receptors are relatively simple in structure and widely distributed in the body

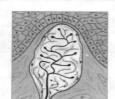

- free nerve endings
- root hair plexus
- tactile discs
- Merkel cells
- tactile corpuscles
- Meissner corpuscles
- lamellated corpuscles
- pacinian corpuscles
- Ruffini corpuscles

13.16

Three major somatic sensory pathways carry information from the skin and musculature to the CNS

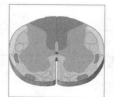

- spinothalamic pathway
- first-order neurons
- second-order neurons
- third-order neurons
- anterior spinothalamic tracts
- lateral spinothalamic tracts
- sensory homunculus
- posterior column pathway
- medial lemniscus
- spinocerebellar pathway

• = *Term boldfaced in this module*

13.17

The somatic nervous system controls skeletal muscles through upper and lower motor neurons

- upper motor neuron
- lower motor neuron
- corticospinal pathway
- motor homunculus
- corticobulbar tracts
- corticospinal tracts
- pyramids
- anterior corticospinal tracts
- lateral corticospinal tracts
- medial pathway
- lateral pathway
- red nucleus
- reticulospinal tracts

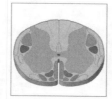

- vestibulospinal tracts
- tectospinal tracts
- rubrospinal tracts

13.18

There are multiple levels of somatic motor control

- brain stem and spinal cord
- pons and medulla oblongata
- hypothalamus
- thalamus and midbrain
- basal nuclei
- cerebral cortex
- cerebellum

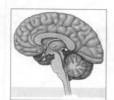

13.19

Nervous system disorders may result from problems with neurons, pathways, or a combination of the two

- referred pain
- Parkinson disease
- rabies
- cerebral palsy (CP)
- amyotrophic lateral sclerosis (ALS)
- Alzheimer disease (AD)
- senile dementia
- multiple sclerosis (MS)

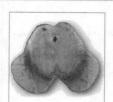

• = *Term boldfaced in this module*

1. Concept map

Use each of the following terms once to fill in the blank spaces to correctly complete the autonomic nervous system concept map.

- sympathetic division
- craniosacral division
- cranial nerves III, VII, IX, X
- thoracolumbar division
- parasympathetic division
- sacral nerves
- thoracic nerves
- enteric nervous system
- lumbar nerves

2. Labeling

Fill in the missing labels in this diagram of the sympathetic division of the ANS. Also indicate the distribution of the sympathetic innervation using red for the preganglionic fibers and black for the postganglionic fibers.

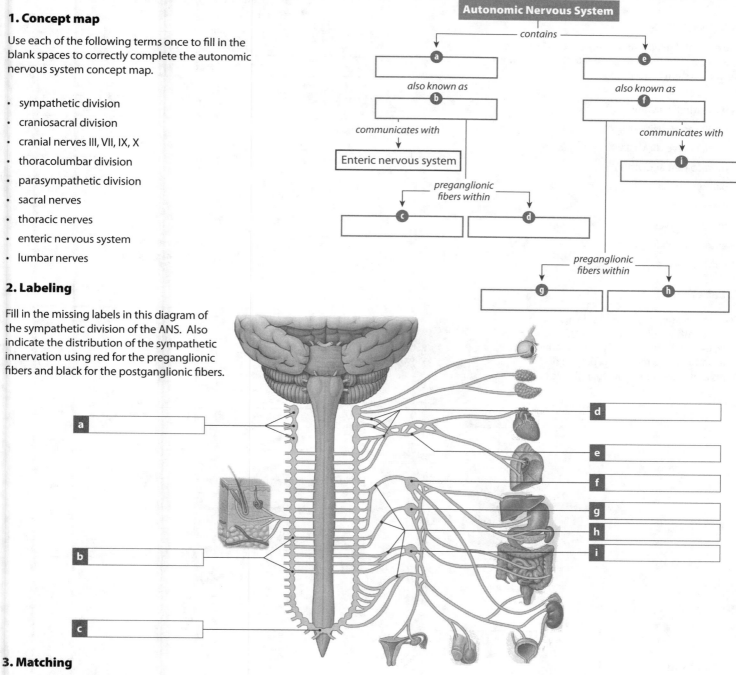

Autonomic Nervous System

contains

a

also known as

b

communicates with

Enteric nervous system

preganglionic fibers within

c   d

e

also known as

f

communicates with

i

preganglionic fibers within

g   h

a
b
c
d
e
f
g
h
i

3. Matching

Match the following terms with the most closely related description.

- nicotinic, muscarinic
- secrete norepinephrine
- receptors
- alpha, beta
- cholinergic
- splanchnic nerves
- acetylcholine
- parasympathetic activation

a	_____	Collateral ganglia
b	_____	Parasympathetic neurotransmitter
c	_____	Sexual arousal
d	_____	Adrenal medullae
e	_____	Cholinergic receptors
f	_____	Determines neurotransmitter effects
g	_____	All parasympathetic neurons
h	_____	Adrenergic receptors

1. Concept map

Use each of the following terms once to fill in the blank spaces to correctly complete the autonomic nervous system concept map.

- sympathetic division
- craniosacral division
- cranial nerves III, VII, IX, X
- thoracolumbar division
- parasympathetic division
- sacral nerves
- thoracic nerves
- enteric nervous system
- lumbar nerves

2. Labeling

Fill in the missing labels in this diagram of the sympathetic division of the ANS. Also indicate the distribution of the sympathetic innervation using red for the preganglionic fibers and black for the postganglionic fibers.

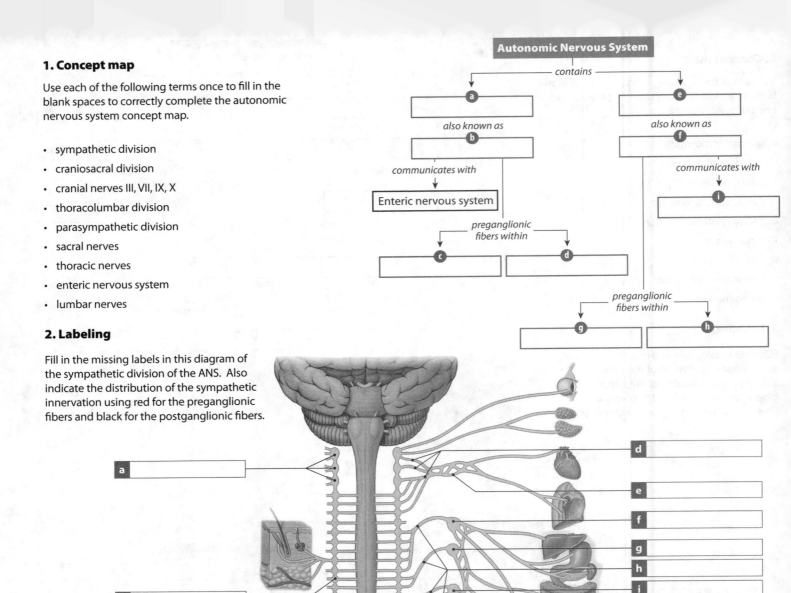

3. Matching

Match the following terms with the most closely related description.

- nicotinic, muscarinic
- secrete norepinephrine
- receptors
- alpha, beta
- cholinergic
- splanchnic nerves
- acetylcholine
- parasympathetic activation

a	_____	Collateral ganglia
b	_____	Parasympathetic neurotransmitter
c	_____	Sexual arousal
d	_____	Adrenal medullae
e	_____	Cholinergic receptors
f	_____	Determines neurotransmitter effects
g	_____	All parasympathetic neurons
h	_____	Adrenergic receptors

1. Concept map

Use each of the following terms once to fill in the blank spaces to correctly complete the levels of autonomic control concept map.

- respiratory
- pons
- spinal cord T$_1$–L$_2$
- vasomotor
- coughing
- hypothalamus
- sympathetic visceral reflexes
- parasympathetic visceral reflexes
- complex visceral reflexes
- limbic system and thalamus

Levels of Autonomic Control

```
                           Cerebral cortex
Communication          ┌──────────a──────────┐ — processes →  Emotions and
at subconscious level  └─────────────────────┘                sensory input
                       ┌──────────b──────────┐ — controls →   Sympathetic and
                       └─────────────────────┘                parasympathetic divisions
                          ┌───────c──────────┐ — function →   Higher levels of
                          └──────────────────┘                respiratory control
                       Medulla oblongata  — processes →   ┌──────e──────┐
                            centers
         ┌──────────┬──────────┬──────────┬──────────┬──────────┐
       Cardiac    ┌──f──┐   Swallowing  ┌──g──┐    ┌──h──┐
       ┌──────d──────┐  — neurons control →  ┌──────i──────┐
       Sacral spinal cord  — neurons control →  ┌──────j──────┐ — such as →  Defecation and urination
```

2. Matching

Write S (for sympathetic) or P (for parasympathetic) to indicate the ANS division responsible for each of the following effects.

a   _____ decreased metabolic rate

b   _____ increased salivary and digestive secretions

c   _____ increased metabolic rate

d   _____ stimulation of urination and defecation

e   _____ activation of sweat glands

f   _____ heightened mental alertness

g   _____ decreased heart rate and blood pressure

h   _____ activation of energy reserves

i   _____ increased heart rate and blood pressure

j   _____ reduced digestive and urinary functions

k   _____ increased motility and blood flow in the digestive tract

l   _____ increased respiratory rate and dilation of respiratory passages

m   _____ constriction of the pupils and focus of the eyes on nearby objects

3. Section integration

Recent surveys show that about one-third of the American adult population is involved in some type of exercise program. What contributions does sympathetic activation make to help the body adjust to changes that occur during exercise and still maintain homeostasis?

_____

_____

1. Concept map

Use each of the following terms once to fill in the blank spaces to correctly complete the levels of autonomic control concept map.

- respiratory
- pons
- spinal cord T$_1$–L$_2$
- vasomotor
- coughing
- hypothalamus
- sympathetic visceral reflexes
- parasympathetic visceral reflexes
- complex visceral reflexes
- limbic system and thalamus

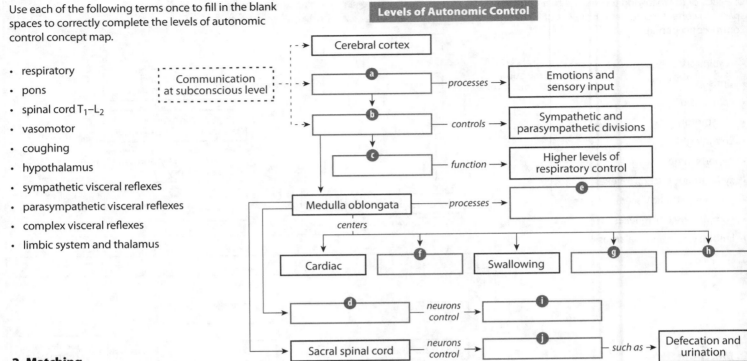

2. Matching

Write S (for sympathetic) or P (for parasympathetic) to indicate the ANS division responsible for each of the following effects.

a _____ decreased metabolic rate

b _____ increased salivary and digestive secretions

c _____ increased metabolic rate

d _____ stimulation of urination and defecation

e _____ activation of sweat glands

f _____ heightened mental alertness

g _____ decreased heart rate and blood pressure

h _____ activation of energy reserves

i _____ increased heart rate and blood pressure

j _____ reduced digestive and urinary functions

k _____ increased motility and blood flow in the digestive tract

l _____ increased respiratory rate and dilation of respiratory passages

m _____ constriction of the pupils and focus of the eyes on nearby objects

3. Section integration

Recent surveys show that about one-third of the American adult population is involved in some type of exercise program. What contributions does sympathetic activation make to help the body adjust to changes that occur during exercise and still maintain homeostasis?

Visual Outline with Key Terms

Summarize the content of each module using the terms in the order provided.

The Functional Anatomy and Organization of the Autonomic Nervous System (ANS)

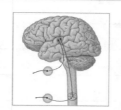

- somatic nervous system (SNS)
- autonomic nervous system (ANS)
- preganglionic neurons
- ganglionic neurons
- autonomic ganglia

_____

_____

_____

_____

_____

_____

_____

_____

_____

14.1

The ANS consists of sympathetic, parasympathetic, and enteric divisions

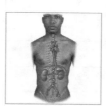

- sympathetic division
- parasympathetic division
- thoracolumbar division
- craniosacral division
- enteric nervous system (ENS)

_____

_____

_____

_____

_____

_____

_____

_____

_____

14.2

The sympathetic division has chain ganglia, collateral ganglia, and the adrenal medullae; whereas the parasympathetic division has terminal or intramural ganglia

- preganglionic fibers
- postganglionic fibers
- sympathetic chain
- collateral ganglia
- celiac ganglia
- superior mesenteric ganglia
- inferior mesenteric ganglia
- adrenal medulla
- terminal ganglia
- intramural ganglia
- ciliary ganglion
- pterygopalatine ganglion
- submandibular ganglion
- otic ganglion

_____

_____

_____

_____

_____

_____

_____

_____

_____

_____

_____

• = _Term boldfaced in this module_

The two ANS divisions innervate many of the same structures, but the innervation patterns are different

- splanchnic nerves
- sympathetic nerves

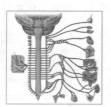

The effects of sympathetic and parasympathetic stimulation are mediated by membrane receptors at the target organs

- adrenergic receptors
- alpha receptors
- beta receptors
- nicotinic receptors
- muscarinic receptors

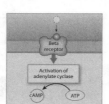

The functional differences between the two ANS divisions reflect their divergent anatomical and physiological characteristics

- sympathetic activation
- parasympathetic stimulation
- anabolic system

● = *Term boldfaced in this module*

Autonomic Regulation and Control Mechanisms

- central nervous system (CNS) processing
- sensory processing centers
- motor centers
- motor responses
- somatic effectors
- visceral effectors
- autonomic output

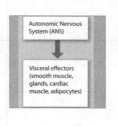

14.6

The ANS provides precise control over visceral functions

- autonomic tone
- dual innervation
  - sympathetic effects
  - parasympathetic effects

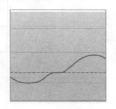

14.7

Most visceral functions are controlled by visceral reflexes

- visceral reflexes
- short reflexes
- long reflexes
- solitary nuclei

• = *Term boldfaced in this module*

14.8

Baroreceptors and chemoreceptors initiate important autonomic reflexes involving visceral sensory pathways

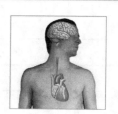

- baroreceptors
- chemoreceptors
- carotid bodies
- aortic bodies

_____

_____

_____
_____
_____
_____
_____
_____

14.9

The autonomic nervous system has multiple levels of motor control

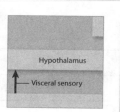

Hypothalamus

Visceral sensory

- cerebral cortex
- limbic system
- thalamus
- hypothalamus
- pons
- medulla oblongata
- spinal cord T_1–L_2
- sacral spinal cord S_2–S_4

_____

_____
_____
_____
_____
_____
_____
_____
_____
_____
_____

• = _Term boldfaced in this module_

1. Matching

Match the following terms with the most closely related description.

- gustation
- depolarization
- Bowman glands
- lingual papillae
- G proteins
- stem cells
- bitter
- olfactory cilia
- olfaction
- taste bud
- cerebral cortex
- olfactory bulb
- odorant

a _____ Sweet, bitter, and umami sensations

b _____ Sense of smell

c _____ Basal cells

d _____ Chemical stimulus

e _____ Sense of taste

f _____ Olfactory glands

g _____ Receives all special senses stimuli

h _____ Cluster of gustatory receptors

i _____ Site of first synapse by olfactory receptors

j _____ Contain olfactory receptor proteins

k _____ Most sensitive taste sensation

l _____ Produces generator potential

m _____ Epithelial projections of tongue

2. Labeling

Label the areas of taste on the tongue (a–e) and the three types of lingual papillae (f–h).

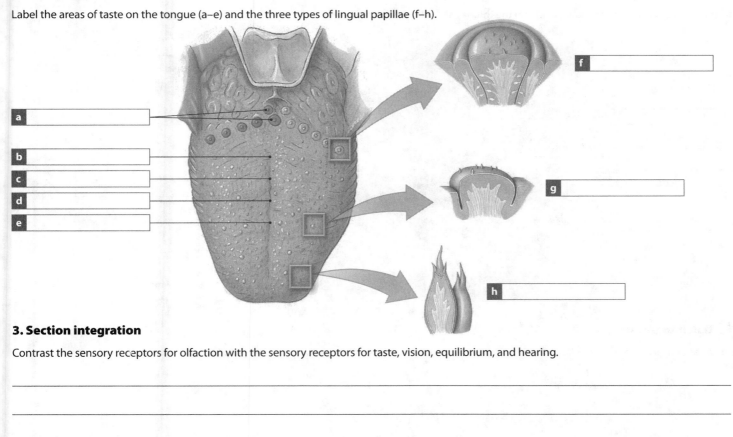

f _____

a _____

b _____

c _____

d _____

e _____

g _____

h _____

3. Section integration

Contrast the sensory receptors for olfaction with the sensory receptors for taste, vision, equilibrium, and hearing.

_____

_____

_____

_____

1. Matching

Match the following terms with the most closely related description.

- gustation
- depolarization
- Bowman glands
- lingual papillae
- G proteins
- stem cells
- bitter
- olfactory cilia
- olfaction
- taste bud
- cerebral cortex
- olfactory bulb
- odorant

a _____ Sweet, bitter, and umami sensations

b _____ Sense of smell

c _____ Basal cells

d _____ Chemical stimulus

e _____ Sense of taste

f _____ Olfactory glands

g _____ Receives all special senses stimuli

h _____ Cluster of gustatory receptors

i _____ Site of first synapse by olfactory receptors

j _____ Contain olfactory receptor proteins

k _____ Most sensitive taste sensation

l _____ Produces generator potential

m _____ Epithelial projections of tongue

2. Labeling

Label the areas of taste on the tongue (a–e) and the three types of lingual papillae (f–h).

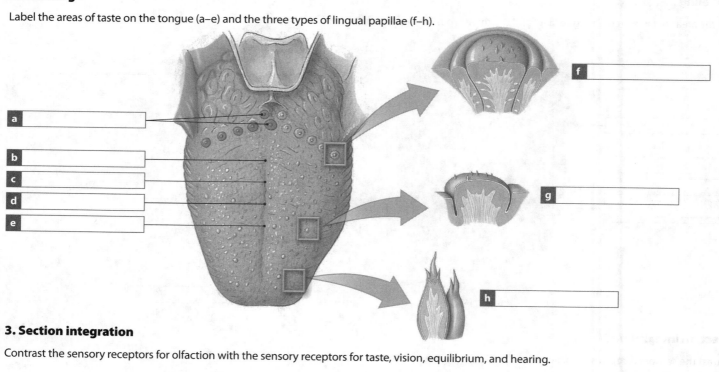

a _____

b _____

c _____

d _____

e _____

f _____

g _____

h _____

3. Section integration

Contrast the sensory receptors for olfaction with the sensory receptors for taste, vision, equilibrium, and hearing.

_____

_____

_____

_____

1. Labeling

Label the structures in the following diagram of the right ear.

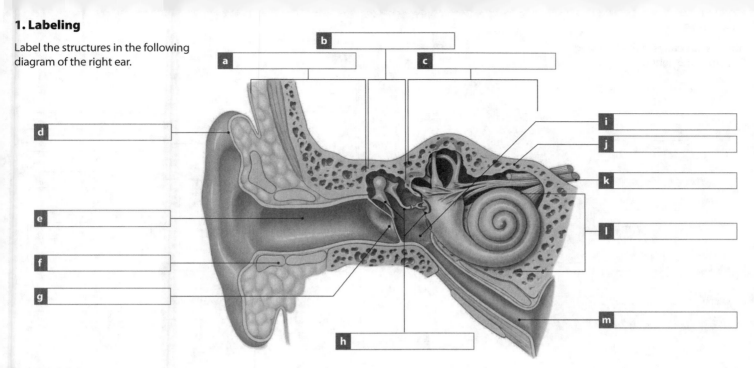

a

b

c

d

e

f

g

h

i

j

k

l

m

2. Matching

Match the following terms with the most closely related description.

- inner ear
- organ of Corti
- round window
- ampulla
- decibel
- auricle
- temporal lobe
- incus
- cerumen
- tectorial membrane
- tympanic membrane
- endolymph
- stapes
- tympanic cavity
- equilibrium

a	_____	Ceruminous glands
b	_____	aids in determining the direction of sound
c	_____	Fluid within chambers and canals of inner ear
d	_____	Opens into auditory tube
e	_____	Membranous labyrinth
f	_____	Utricle, saccule
g	_____	Lies at end of external acoustic meatus
h	_____	Attached to oval window
i	_____	Separates perilymph of cochlea from middle ear
j	_____	Middle auditory ossicle
k	_____	Contains receptors for rotation in semicircular ducts
l	_____	Auditory cortex
m	_____	Cochlear structure that provides information to the CNS
n	_____	Overlies hair cells in the cochlear duct
o	_____	Unit of sound intensity

3. Section integration

For a few seconds after you ride the express elevator from the 25th floor to the ground floor, you still feel as if you are descending, even though you have come to a stop. Why?

_____

_____

1. Labeling

Label the structures in the following diagram of the right ear.

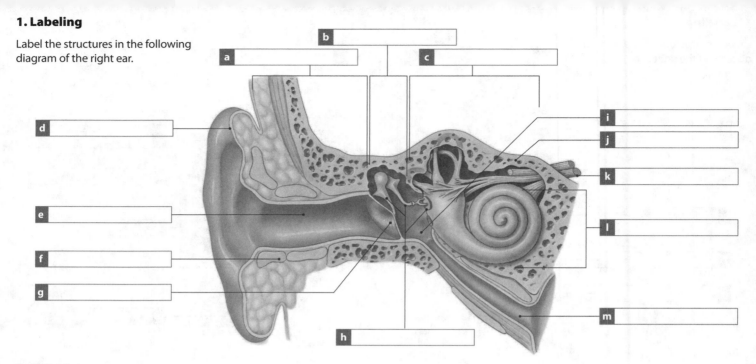

2. Matching

Match the following terms with the most closely related description.

- inner ear
- organ of Corti
- round window
- ampulla
- decibel
- auricle
- temporal lobe
- incus
- cerumen
- tectorial membrane
- tympanic membrane
- endolymph
- stapes
- tympanic cavity
- equilibrium

a	_____	Ceruminous glands
b	_____	aids in determining the direction of sound
c	_____	Fluid within chambers and canals of inner ear
d	_____	Opens into auditory tube
e	_____	Membranous labyrinth
f	_____	Utricle, saccule
g	_____	Lies at end of external acoustic meatus
h	_____	Attached to oval window
i	_____	Separates perilymph of cochlea from middle ear
j	_____	Middle auditory ossicle
k	_____	Contains receptors for rotation in semicircular ducts
l	_____	Auditory cortex
m	_____	Cochlear structure that provides information to the CNS
n	_____	Overlies hair cells in the cochlear duct
o	_____	Unit of sound intensity

3. Section integration

For a few seconds after you ride the express elevator from the 25th floor to the ground floor, you still feel as if you are descending, even though you have come to a stop. Why?

_____

_____

1. Labeling

Label the structures in the following diagram of a sagittal section of the left eye.

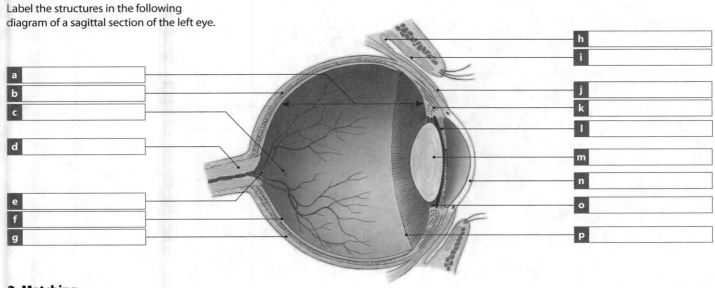

a _____

b _____

c _____

d _____

e _____

f _____

g _____

h _____

i _____

j _____

k _____

l _____

m _____

n _____

o _____

p _____

2. Matching

Match the following terms with the most closely related description.

- ganglion cells
- cones
- vascular tunic
- rods
- optic disc
- posterior chamber
- crystallins
- sclera
- retina
- occipital lobe
- palpebrae
- rhodopsin
- fovea
- posterior cavity
- pupil

a _____ Visual pigment

b _____ Eyelids

c _____ Transparent proteins in the cells of a lens

d _____ White of the eye

e _____ Opening surrounded by the iris

f _____ Neural tunic

g _____ Site of vitreous body

h _____ Photoreceptors that enable vision in dim light

i _____ Extends between the iris and the ciliary body and lens

j _____ Sharpest vision

k _____ Visual cortex

l _____ Photoreceptors that provide the perception of color

m _____ Their axons form the optic nerves

n _____ Iris, ciliary body, and choroid

o _____ Region of retina called the "blind spot"

3. Section integration

A bright flash of light from nearby exploding fireworks blinds Rachel's eyes. The result is a "ghost" image that temporarily remains on her retinas. What might account for the images and their subsequent disappearance?

_____

_____

_____

1. Labeling

Label the structures in the following diagram of a sagittal section of the left eye.

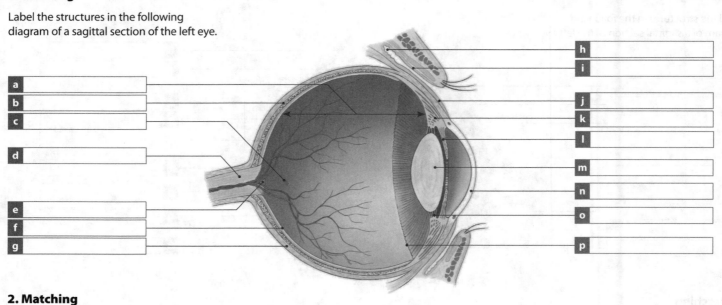

a _____
b _____
c _____
d _____
e _____
f _____
g _____

h _____
i _____
j _____
k _____
l _____
m _____
n _____
o _____
p _____

2. Matching

Match the following terms with the most closely related description.

- ganglion cells
- cones
- vascular tunic
- rods
- optic disc
- posterior chamber
- crystallins
- sclera
- retina
- occipital lobe
- palpebrae
- rhodopsin
- fovea
- posterior cavity
- pupil

a _____ Visual pigment

b _____ Eyelids

c _____ Transparent proteins in the cells of a lens

d _____ White of the eye

e _____ Opening surrounded by the iris

f _____ Neural tunic

g _____ Site of vitreous body

h _____ Photoreceptors that enable vision in dim light

i _____ Extends between the iris and the ciliary body and lens

j _____ Sharpest vision

k _____ Visual cortex

l _____ Photoreceptors that provide the perception of color

m _____ Their axons form the optic nerves

n _____ Iris, ciliary body, and choroid

o _____ Region of retina called the "blind spot"

3. Section integration

A bright flash of light from nearby exploding fireworks blinds Rachel's eyes. The result is a "ghost" image that temporarily remains on her retinas. What might account for the images and their subsequent disappearance?

_____

_____

_____

Visual Outline with Key Terms

Summarize the content of each module using the terms in the order provided.

An Introduction to the Special Senses: Olfaction and Gustation

- generator potential
- synaptic delay

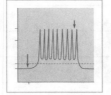

_____

_____

_____

_____

_____

_____

_____

15.1

Olfaction involves specialized chemoreceptive neurons and delivers sensations directly to the cerebrum

- olfactory organs
- olfactory bulb
- olfactory tract
- olfactory epithelium
- lamina propria
- olfactory glands
- olfactory receptor cells
- supporting cells
- basal cells
- odorants

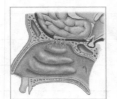

_____

_____

_____

_____

_____

_____

_____

15.2

Gustation involves epithelial chemoreceptor cells located in taste buds

- gustation
- taste receptors
- lingual papillae
- taste buds
- circumvallate papillae
- fungiform papillae
- filiform papillae
- umami
- water receptors
- gustatory cells
- taste pore
- basal cells

_____

_____

_____

_____

_____

_____

_____

● = _Term boldfaced in this module_

15.3

Gustatory reception relies on membrane receptors and channels, and sensations are carried by facial, glossopharyngeal, and vagus nerves

○ central adaptation
• gustducins
• solitary nucleus
○ thalamus
○ primary sensory cortex

SECTION 2

Equilibrium and Hearing

• inner ear
• hair cells
• kinocilium
• stereocilia

15.4

The ear is divided into the external ear, the middle ear, and the inner ear

• external ear
• middle ear
• inner ear
• auricle
• external acoustic meatus
• ceruminous glands
• cerumen

• auditory ossicles
• tympanic membrane
• bony labyrinth
• auditory tube
• otitis media
• malleus
• incus

• stapes
• tensor tympani muscle
• stapedius muscle

• = *Term boldfaced in this module*

164

15.5

The bony labyrinth protects the membranous labyrinth

- inner ear
- bony labyrinth
- perilymph
- semicircular canals
- vestibule
- cochlea
- membranous labyrinth
- endolymph
- semicircular ducts
- utricle
- saccule
- cochlear duct
- vestibular complex

15.6

Hair cells in the semicircular ducts respond to rotation, while those in the utricle and saccule respond to gravity and linear acceleration

- anterior semicircular duct
- posterior semicircular duct
- lateral semicircular duct
- ampulla
- crista
- cupula
- endolymphatic duct
- endolymphatic sac
- statoconia
- otolith
- maculae

15.7

The cochlear duct contains the hair cells of the organ of Corti

- cochlear duct
- vestibular duct
- tympanic duct
- oval window
- round window
- vestibular membrane
- basilar membrane
- organ of Corti
- spiral ganglion
- tectorial membrane

• = *Term boldfaced in this module*

15.8

The organ of Corti provides sensations of pitch and volume

- wavelength
- frequency
- hertz (Hz)
- amplitude
- intensity
- decibels (dB)
- resonance
- pitch
- volume

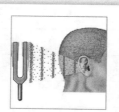

15.9

The vestibulocochlear nerve carries equilibrium and hearing sensations to the brain stem

- vestibular ganglion
- vestibular nuclei
- spiral ganglion
- cochlear nuclei
- inferior colliculus
- thalamus
- auditory cortex

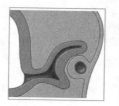

SECTION 3

Vision

- optic vesicles
- optic cups
- ependymal cells

• = *Term boldfaced in this module*

15.10

The accessory structures of the eye provide protection while allowing light to reach the interior of the eye

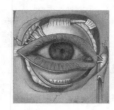

- accessory structures
- cornea
- lateral canthus
- pupil
- iris
- lacrimal caruncle
- eyelashes
- palpebra
- medial canthus
- palpebral fissure
- conjunctiva
- tarsal glands
- palpebral conjunctiva
- ocular conjunctiva
- fornix
- lacrimal apparatus
- lacrimal gland
- lysozyme
- tear ducts
- lacrimal puncta
- lacrimal canaliculi
- lacrimal sac
- nasolacrimal duct
- conjunctivitis

15.11

The eye has a layered wall; it is hollow, with fluid-filled anterior and posterior cavities

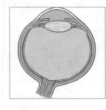

- tunics
- fibrous tunic
- sclera
- corneal limbus
- vascular tunic (uvea)
- iris
- ciliary body
- choroid
- neural tunic (retina)
- photoreceptors
- anterior cavity
- anterior chamber
- posterior chamber
- posterior cavity
- vitreous body
- ciliary muscle
- ciliary processes
- ora serrata
- suspensory ligaments
- aqueous humor
- canal of Schlemm

15.12

The eye is highly organized and has a consistent visual axis that directs light to the fovea of the retina

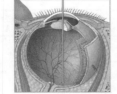

- crystallins
- optic nerve
- pupillary dilator muscles
- papillary constrictor muscles
- visual axis
- macula lutea
- fovea

● = *Term boldfaced in this module*

15.13

Focusing produces a sharply defined image at the retina

- refracted
- focal point
- focal distance
- accommodation
- near point of vision

15.14

The neural tissue of the retina contains multiple layers of specialized photoreceptors, neurons, and supporting cells

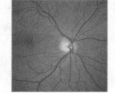

- ganglion cells
- optic disc
- blind spot
- rods
- cones
- bipolar cells
- horizontal cells
- amacrine cells
- visual acuity

15.15

Photoreception, which occurs in the outer segment of rods and cones, involves the activation of visual pigments

- visual pigments
- pigment epithelium
- photoreceptor
- outer segment
- discs
- inner segment
- rhodopsin
- opsin
- retinal
- bleaching
- blue cones
- green cones
- red cones
- color blindness

• = *Term boldfaced in this module*

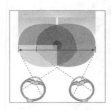

1. Concept map

Place each of the following terms in the appropriate box to correctly complete the hormones concept map.

- steroid hormones
- tryptophan derivatives
- glycoproteins
- short polypeptides
- catecholamines
- peptide hormones
- thyroid hormones
- transport proteins
- lipid derivatives
- small proteins
- eicosanoids

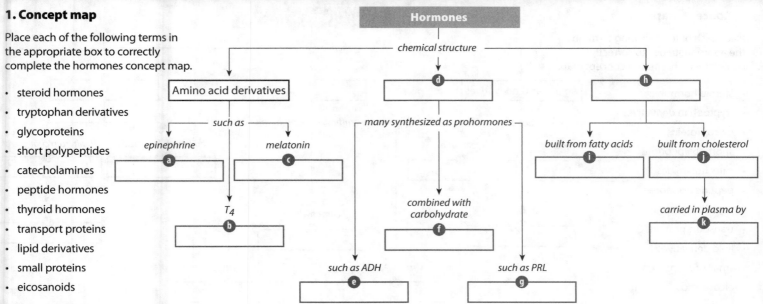

2. Short answer

Identify the endocrine gland—or organ containing endocrine cells—based on the major effects produced by their secreted hormone(s).

a _____ stimulates and coordinates the immune response

b _____ establishes day/night cycles

c _____ secretes insulin and regulates glucose uptake and utilization

d _____ controls hormone secretion of the pituitary gland

e _____ regulates red blood cell production and the absorption of calcium and phosphate by the intestinal tract

f _____ regulates mineral balance, metabolic control, and resistance to stress

g _____ regulates secretions of adrenal cortex, thyroid gland, and reproductive organs

h _____ affects growth, metabolism, and sexual characteristics

i _____ maintains circulation and has a role in regulating blood volume

j _____ coordinates its system functions, glucose metabolism, and appetite

k _____ regulates metabolic rate and calcium levels in body fluids

l _____ glands embedded in posterior surface of thyroid that play an important role in the response to decreasing calcium levels in body fluids

3. Matching

Match the following terms with the most closely related description.

- FSH
- androgens
- F cells
- parathyroid glands
- epinephrine
- direct communication
- tropic hormones
- secretes releasing hormones
- prostaglandins
- cyclic-AMP

a _____ pancreatic polypeptide

b _____ adrenal medulla

c _____ gap junctions

d _____ pituitary gland

e _____ second messenger

f _____ hypothalamus

g _____ zona reticularis

h _____ eicosanoids

i _____ gonadotropin

j _____ chief cells

1. Concept map

Place each of the following terms in the appropriate box to correctly complete the hormones concept map.

- steroid hormones
- tryptophan derivatives
- glycoproteins
- short polypeptides
- catecholamines
- peptide hormones
- thyroid hormones
- transport proteins
- lipid derivatives
- small proteins
- eicosanoids

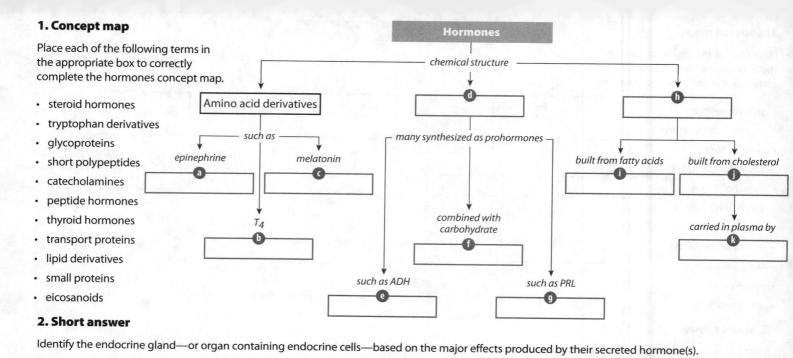

2. Short answer

Identify the endocrine gland—or organ containing endocrine cells—based on the major effects produced by their secreted hormone(s).

a	_____	stimulates and coordinates the immune response
b	_____	establishes day/night cycles
c	_____	secretes insulin and regulates glucose uptake and utilization
d	_____	controls hormone secretion of the pituitary gland
e	_____	regulates red blood cell production and the absorption of calcium and phosphate by the intestinal tract
f	_____	regulates mineral balance, metabolic control, and resistance to stress
g	_____	regulates secretions of adrenal cortex, thyroid gland, and reproductive organs
h	_____	affects growth, metabolism, and sexual characteristics
i	_____	maintains circulation and has a role in regulating blood volume
j	_____	coordinates its system functions, glucose metabolism, and appetite
k	_____	regulates metabolic rate and calcium levels in body fluids
l	_____	glands embedded in posterior surface of thyroid that play an important role in the response to decreasing calcium levels in body fluids

3. Matching

Match the following terms with the most closely related description.

- FSH
- androgens
- F cells
- parathyroid glands
- epinephrine
- direct communication
- tropic hormones
- secretes releasing hormones
- prostaglandins
- cyclic-AMP

a	_____	pancreatic polypeptide
b	_____	adrenal medulla
c	_____	gap junctions
d	_____	pituitary gland
e	_____	second messenger
f	_____	hypothalamus
g	_____	zona reticularis
h	_____	eicosanoids
i	_____	gonadotropin
j	_____	chief cells

1. Matching

Match the following terms with the most closely related description.

- sympathetic activation
- reduce blood pressure and volume
- PTH and calcitonin
- GH and glucocorticoids
- increase blood pressure and volume
- homeostasis threat
- glucocorticoids
- PTH and calcitriol
- gigantism
- protein synthesis

a _____ antagonistic effect

b _____ resistance phase

c _____ alarm phase

d _____ renin and EPO effect

e _____ growth hormone (GH)

f _____ additive effect

g _____ excessive GH in children

h _____ natriuretic peptides effect

i _____ stress

j _____ integrative effect

2. Short answer

Write the phrases at left in the boxes to complete the diagram of the homeostatic regulation of blood pressure and volume.

- increased fluid loss
- erythropoietin released
- reduced blood pressure
- aldosterone secreted
- suppression of thirst
- falling blood pressure and volume
- rising blood pressure and volume
- renin released
- ADH secreted
- release of natriuretic peptides
- Na+ and H$_2$O loss from kidneys
- increased red blood cell production

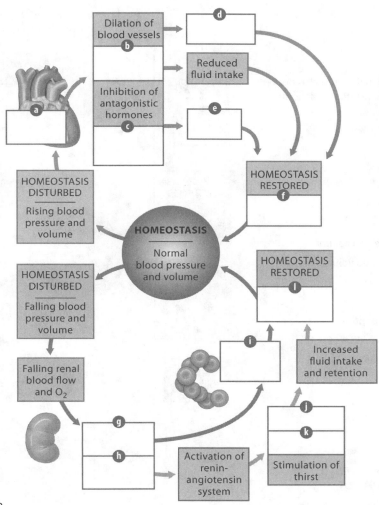

3. Section integration

Describe, and give an example of, the four possible effects that may occur when a cell receives instructions from two different hormones.

_____

_____

1. Matching

Match the following terms with the most closely related description.

- sympathetic activation
- reduce blood pressure and volume
- PTH and calcitonin
- GH and glucocorticoids
- increase blood pressure and volume
- homeostasis threat
- glucocorticoids
- PTH and calcitriol
- gigantism
- protein synthesis

a	_____	antagonistic effect
b	_____	resistance phase
c	_____	alarm phase
d	_____	renin and EPO effect
e	_____	growth hormone (GH)
f	_____	additive effect
g	_____	excessive GH in children
h	_____	natriuretic peptides effect
i	_____	stress
j	_____	integrative effect

2. Short answer

Write the phrases at left in the boxes to complete the diagram of the homeostatic regulation of blood pressure and volume.

- increased fluid loss
- erythropoietin released
- reduced blood pressure
- aldosterone secreted
- suppression of thirst
- falling blood pressure and volume
- rising blood pressure and volume
- renin released
- ADH secreted
- release of natriuretic peptides
- Na$^+$ and H$_2$O loss from kidneys
- increased red blood cell production

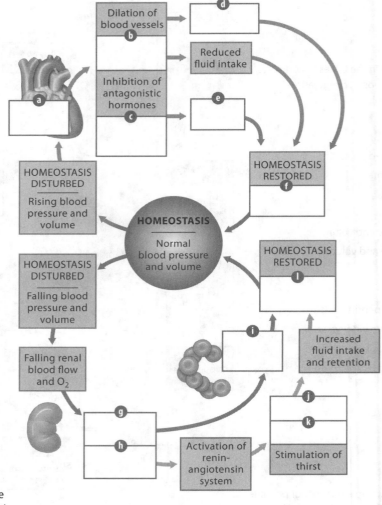

3. Section integration

Describe, and give an example of, the four possible effects that may occur when a cell receives instructions from two different hormones.

Visual Outline with Key Terms

Summarize the content of each module using the terms in the order provided.

SECTION 1

Hormones and Intercellular Communication

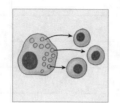

- direct communication
- paracrine communication
- **paracrine factors**
- endocrine communication
- **hormones**
- synaptic communication
- neurotransmitters

_____

_____

_____

_____

_____

_____

_____

_____

16.1

Hormones may be amino acid derivatives, peptides, or lipid derivatives

Catecholamines

Epinephrine

- **amino acid derivatives**
- **peptide hormones**
- **lipid derivatives**
- **catecholamines**
- **prohormones**
- **eicosanoids**
- **leukotrienes**
- **prostaglandins**
- **steroid hormones**
- **endocrine system**
- **hypothalamus**
- **pituitary gland**
- **thyroid gland**
- **adrenal glands**
- **pancreatic islets**
- **pineal gland**
- **parathyroid glands**

_____

_____

_____

_____

_____

_____

_____

_____

16.2

Hormones affect target cells after binding to receptors in the plasma membrane, cytoplasm, or nucleus

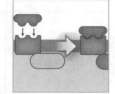

- **first messenger**
- **second messenger**
- **cyclic-AMP (cAMP)**
- calcium ions
- **G protein**
- **calmodulin**
- steroid hormones
- thyroid hormones

_____

_____

_____

_____

_____

_____

_____

_____

• = _Term boldfaced in this module_

16.3

The hypothalamus exerts direct or indirect control over the activities of many different endocrine organs

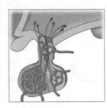

- hypothalamus
- regulatory hormones
- median eminence
- fenestrated capillaries
- hypophyseal portal system
- portal vessels
- antidiuretic hormone (ADH)
- oxytocin (OXT)
- releasing hormones (RH)
- inhibiting hormones (IH)

16.4

The pituitary gland consists of an anterior lobe and a posterior lobe

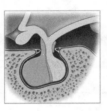

- hypophysis
- tropic hormones
- infundibulum
- adenohypophysis
- neurohypophysis
- antidiuretic hormone (ADH)
- arginine vasopressin (AVP)
- osmoreceptors
- oxytocin (OXT)
- neuroendocrine reflex
- thyroid-stimulating hormone (TSH)
- thyrotropin-releasing hormone (TRH)
- adrenocorticotropic hormone (ACTH)
- corticotropin-releasing hormone (CRH)
- gonadotropins
- gonadotropin-releasing hormone (GnRH)
- follicle-stimulating hormone (FSH)
- estrogens
- inhibin
- luteinizing hormone (LH)
- androgens
- growth hormone (GH)
- growth hormone–releasing hormone (GH–RH)
- growth hormone–inhibiting hormone (GH–IH)
- somatomedins
- glucose-sparing effect
- prolactin (PRL)
- prolactin-inhibiting hormone (PIH)
- melanocyte-stimulating hormone (MSH)

• = Term boldfaced in this module

16.5

Negative feedback mechanisms control the secretion rates of the hypothalamus and pituitary gland

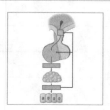

○ negative feedback
● somatomedins

16.6

The thyroid gland contains follicles and requires iodine to produce hormones that stimulate tissue metabolism

- thyroid gland
- isthmus
- thyroid follicles
- thyroglobulin
- C (clear) cells
- calcitonin (CT)
- thyroid hormones (T_4 and T_3)
- thyroid-binding globulins (TBGs)

16.7

Parathyroid hormone, produced by the parathyroid glands, is the primary regulator of calcium ion levels in body fluids

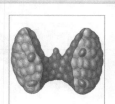

- parathyroid glands
- parathyroid cells
- parathyroid hormone (PTH)
○ calcitriol
○ kidneys

● = *Term boldfaced in this module*

16.8

The adrenal glands produce hormones involved in metabolic regulation

- adrenal gland
- retroperitoneal
- adrenal cortex
- adrenocortical steroids
- corticosteroids
- zona glomerulosa
- mineralocorticoids (MCs)
- aldosterone

- zona fasciculata
- glucocorticoids (GCs)
- cortisol
- corticosterone
- cortisone
- glucose-sparing effect
- anti-inflammatory effect

- zona reticularis
- androgens
- adrenal medulla
- epinephrine (E)
- norepinephrine (NE)

16.9

The pancreatic islets secrete insulin and glucagon and regulate glucose utilization by most cells

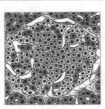

- pancreas
- exocrine pancreas
- endocrine pancreas
- pancreatic islets
- alpha cells
- glucagon

- beta cells
- insulin
- pancreatic acini
- delta cells
- F cells
- GH-IH

- pancreatic polypeptide (PP)

16.10

The pineal gland of the epithalamus secretes melatonin

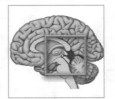

- pineal gland
- pinealocytes
- melatonin
- free radicals
- circadian rhythms

• = *Term boldfaced in this module*

16.11

Diabetes mellitus is an endocrine disorder characterized by excessively high blood glucose levels

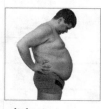

- diabetes mellitus
- hyperglycemia
- glycosuria
- polyuria
- type 1 (insulin dependent) diabetes
- type 2 (non-insulin dependent) diabetes
- diabetic retinopathy
- diabetic nephropathy
- diabetic neuropathy

SECTION 2

Hormones and System Integration

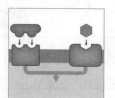

- antagonistic effects
- additive effects
- synergistic effect
- permissive effects
- integrative effects

16.12

Long-term regulation of blood pressure, blood volume, and growth involves hormones produced by the endocrine system and by endocrine tissues in other systems

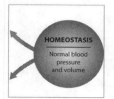

- natriuretic peptides
- erythropoietin (EPO)
- renin
- renin-angiotensin system
- insulin
- parathyroid hormone
- calcitriol
- thyroid hormones
- reproductive hormones
- growth hormone

16.13

The stress response is a predictable response to any significant threat to homeostasis

- stress
- stress response
- general adaptation syndrome (GAS)
- alarm phase
- resistance phase
- exhaustion phase

16.14

Endocrine disorders may result from overproduction or underproduction of hormones

- hypersecretion
- hyposecretion
- acromegaly
- goiter
- cretinism
- Addison disease
- Cushing disease

 = *Term boldfaced in this module*

1. Concept map

Use the terms in the following list to complete the whole blood concept map.

- globulins
- electrolytes, glucose, urea
- protein
- albumins
- leukocytes
- fibrinogen
- solutes
- basophils
- erythrocytes
- formed elements
- plasma
- eosinophils
- lymphocytes
- water
- platelets
- monocytes
- neutrophils

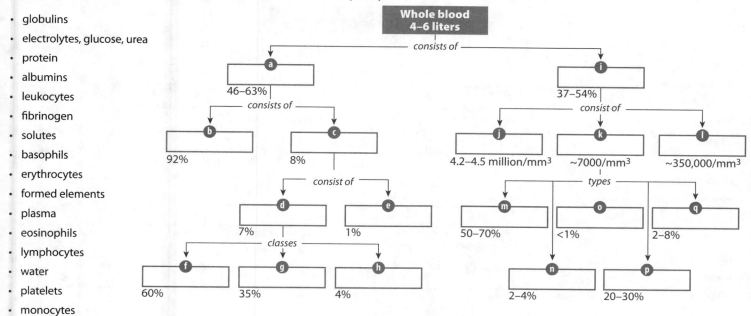

2. Matching

Match the following terms with the most closely related description.

- matrix
- transport protein
- jaundice
- venipuncture
- red bone marrow
- mature RBCs
- pigment complex
- platelets
- erythropoietin
- cross-reaction
- monocytes
- lymphocytes

a	_____	myeloid tissue
b	_____	anucleated
c	_____	plasma
d	_____	macrophages
e	_____	globulin
f	_____	agglutination
g	_____	specific immunity
h	_____	bilirubin
i	_____	median cubital vein
j	_____	heme
k	_____	hormone
l	_____	blood clotting

3. Section integration

Why are mature red blood cells in humans incapable of protein synthesis and mitosis?

_____

_____

1. Concept map

Use the terms in the following list to complete the whole blood concept map.

- globulins
- electrolytes, glucose, urea
- protein
- albumins
- leukocytes
- fibrinogen
- solutes
- basophils
- erythrocytes
- formed elements
- plasma
- eosinophils
- lymphocytes
- water
- platelets
- monocytes
- neutrophils

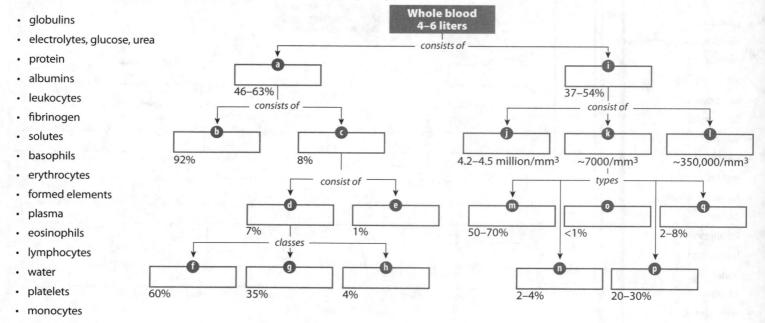

2. Matching

Match the following terms with the most closely related description.

- matrix
- transport protein
- jaundice
- venipuncture
- red bone marrow
- mature RBCs
- pigment complex
- platelets
- erythropoietin
- cross-reaction
- monocytes
- lymphocytes

a	_____	myeloid tissue
b	_____	anucleated
c	_____	plasma
d	_____	macrophages
e	_____	globulin
f	_____	agglutination
g	_____	specific immunity
h	_____	bilirubin
i	_____	median cubital vein
j	_____	heme
k	_____	hormone
l	_____	blood clotting

3. Section integration

Why are mature red blood cells in humans incapable of protein synthesis and mitosis?

_____

_____

1. Labeling

Label the major arteries in the diagram at right.

a

b

c

d

e

f

g

h

i

j

k

l

m

n

2. Labeling

Label the major veins in the accompanying diagram.

a

b

c

d

e

f

g

h

i

j

k

l

m

n

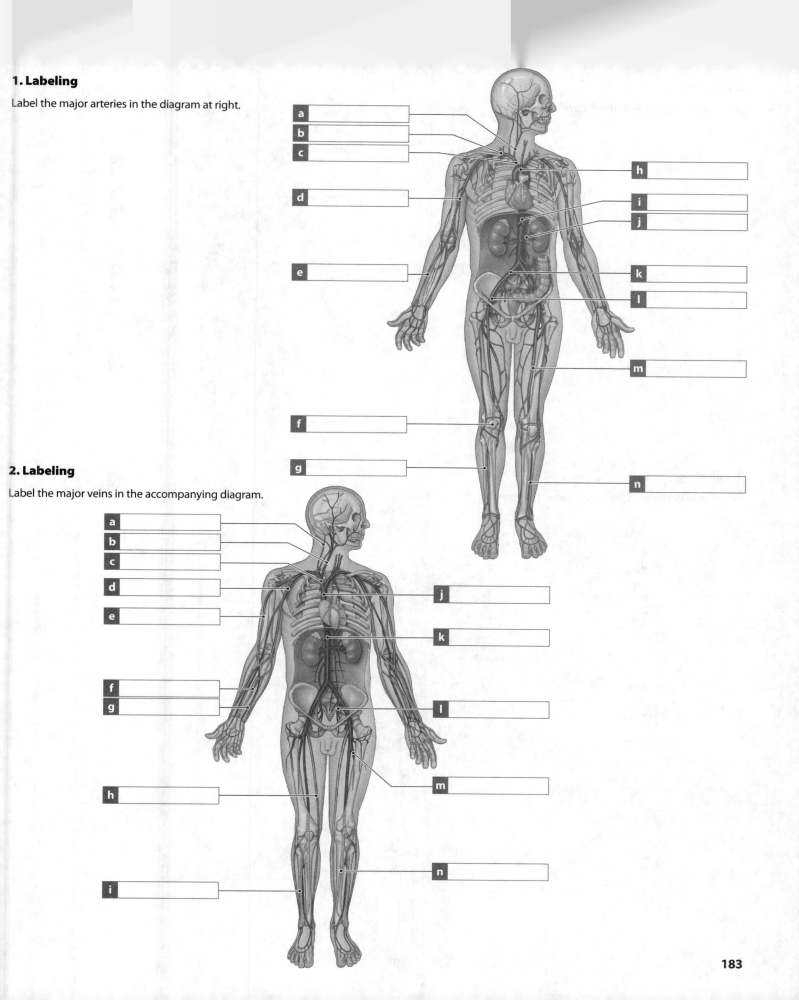

1. Labeling

Label the major arteries in the diagram at right.

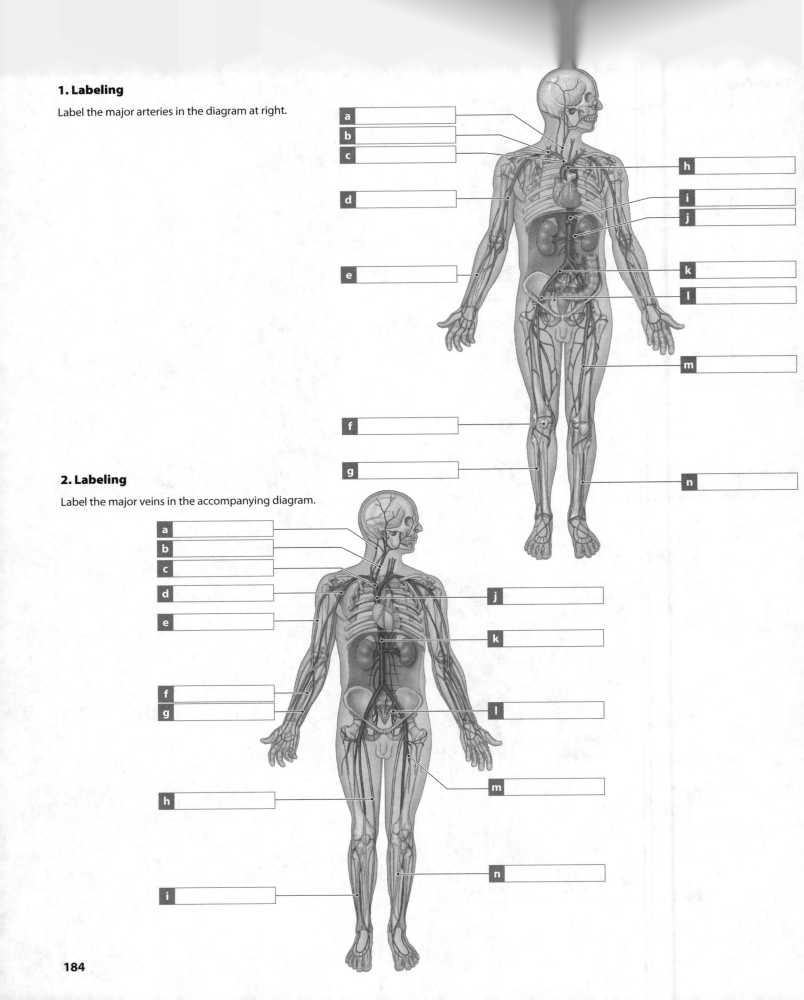

a

b

c

d

e

f

g

h

i

j

k

l

m

n

2. Labeling

Label the major veins in the accompanying diagram.

a

b

c

d

e

f

g

h

i

j

k

l

m

n

Visual Outline with Key Terms

Summarize the content of each module using the terms in the order provided.

Blood

- cardiovascular system
  - heart
  - blood vessels
  - blood

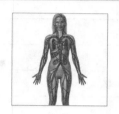

_____

17.1

Blood is a fluid connective tissue containing plasma and formed elements

- plasma
- formed elements
- whole blood
- albumin
- globulins
- immunoglobulins
- transport globulins
- fibrinogen
- fibrin
- electrolytes
- organic nutrients
- organic wastes
- platelets
- white blood cells (WBCs)
- red blood cells (RBCs)
- hematocrit
- packed cell volume (PCV)

_____

17.2

Red blood cells, the most common formed elements, contain hemoglobin

- red blood cell count
- rouleaux
- hemoglobin (Hb)
- alpha (α) chains
- beta (β) chains
- heme
- oxyhemoglobin (HbO$_2$)
- deoxyhemoglobin

_____

• = *Term boldfaced in this module*

17.3

Red blood cells are continuously produced and recycled

- proerythroblasts
- erythroblasts
- reticulocyte
- erythropoiesis
- myeloid tissue
- hemolysis
- hemoglobinuria
- hemolyze
- transferrin
- biliverdin
- bilirubin
- jaundice
- urobilins
- stercobilins
- hematuria

17.4

Blood type is determined by the presence or absence of specific surface antigens on RBCs

- antigens
- immune response
- surface antigens
- blood type
- Type A
- Type B
- Type AB
- Type O
- agglutinate
- agglutination
- cross-reaction
- Rh positive (Rh⁺)
- Rh negative (Rh⁻)
- transfusion reactions
- compatible

17.5

Hemolytic disease of the newborn is an RBC-related disorder caused by a cross-reaction between fetal and maternal blood types

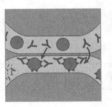

- hemolytic disease of the newborn (HDN)
- sensitization
- erythroblastosis fetalis

• = *Term boldfaced in this module*

17.6

White blood cells defend the body against pathogens, toxins, cellular debris, and abnormal or damaged cells

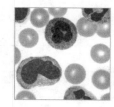

- white blood cells (WBCs)
- leukocytes
- emigration (diapedesis)
- positive chemotaxis
- granular leukocytes
- neutrophils
- eosinophils
- basophils
- agranular leukocytes
- monocytes
- lymphocytes
- differential count

17.7

Formed elements are produced by stem cells derived from hemocytoblasts

- hemocytoblasts
- lymphoid stem cells
- lymphoid tissues
- colony-stimulating factors (CSFs)
- blast cells
- myeloid stem cells
- progenitor cells
- myelocytes
- band cells
- megakaryocytes
- erythropoietin (EPO)
- hypoxia
- platelets

17.8

The clotting response is a complex cascade of events that reduces blood loss

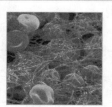

- hemostasis
- vascular phase
- endothelins
- platelet phase
- platelet factors
- platelet-derived growth factor (PDGF)
- coagulation phase
- coagulation
- ○ fibrinogen
- fibrin
- procoagulants
- cascade
- common pathway
- prothrombinase
- prothrombin
- thrombin
- extrinsic pathway
- tissue factor
- intrinsic pathway
- PF-3
- clot retraction
- fibrinolysis
- plasminogen
- tissue plasminogen activator (t-PA)
- plasmin

● = *Term boldfaced in this module*

17.9

Blood disorders can be classified by their origins and the changes in blood characteristics

- venipuncture
- iron deficiency anemia
- microcytic
- vitamin B$_{12}$
- pernicious anemia
- macrocytic
- vitamin K
- sickle cell anemia
- sickling trait
- hemophilia
- thalassemias
- bacteremia
- viremia
- sepsis
- septicemia
- malaria
- leukemias
- myeloid leukemia
- lymphoid leukemia
- disseminated intravascular coagulation (DIC)

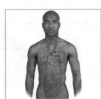

SECTION 2

The Functional Anatomy of Blood Vessels

- pulmonary circuit
- systemic circuit
- arteries
- veins
- capillaries
- right atrium
- right ventricle
- left atrium
- left ventricle

17.10

Arteries and veins differ in the structure and thickness of their walls

- tunica intima
- internal elastic membrane
- tunica media
- vasoconstriction
- vasodilation
- tunica externa
- arteries
- veins
- large veins
- medium-sized veins
- venules
- elastic arteries
- muscular arteries
- arterioles
- capillaries

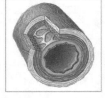

• = *Term boldfaced in this module*

17.11

Capillary structure and capillary blood flow affect the rates of exchange between the blood and interstitial fluid

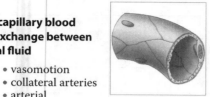

- continuous capillary
- fenestrated capillary
- sinusoids
- capillary bed
- metarteriole
- thoroughfare channel
- vasomotion
- collateral arteries
- arterial anastomosis
- precapillary sphincter
- arteriovenous anastomosis

17.12

The venous system has low pressures and contains almost two-thirds of the body's blood volume

- valves
- varicose veins
- hemorrhoids
- venoconstriction

17.13

The pulmonary circuit, which is relatively short, carries deoxygenated blood from the right ventricle to the lungs and returns oxygenated blood to the left atrium

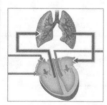

- occlusion
- trunks
- pulmonary trunk
- pulmonary arteries
- pulmonary arterioles
- alveolar capillaries
- alveoli
- pulmonary veins

● = *Term boldfaced in this module*

17.14

The systemic arterial and venous systems operate in parallel, and the major vessels often have similar names

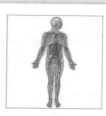

- arterial system
- venous system
- superior vena cava
- inferior vena cava
- dual venous drainage

_____

_____

_____

_____

_____

_____

_____

_____

_____

_____

_____

17.15

The branches of the aortic arch supply structures that are drained by the superior vena cava

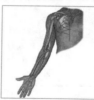

- brachiocephalic trunk
- right subclavian artery
- right common carotid artery
- left common carotid artery
- left subclavian artery
- internal thoracic artery
- vertebral artery
- axillary artery
- brachial artery
- radial artery
- ulnar artery
- palmar arches
- digital arteries
- digital veins
- superficial palmar arch
- median antebrachial vein
- deep palmar arch
- radial vein
- ulnar vein
- median cubital vein
- cephalic vein
- axillary vein
- basilic vein
- brachial vein
- subclavian vein
- external jugular vein
- internal jugular vein
- vertebral vein
- brachiocephalic vein
- superior vena cava (SVC)
- internal thoracic vein

_____

_____

_____

_____

_____

_____

_____

_____

_____

_____

_____

_____

_____

_____

_____

_____

_____

_____

_____

_____

17.16

The external carotid arteries supply the neck, lower jaw, and face, and the internal carotid and vertebral arteries supply the brain, while the external jugular veins drain the regions supplied by the external carotids, and the internal jugular veins drain the brain

- common carotid arteries
- internal carotid artery
- vertebral artery
- basilar artery
- external carotid artery
- carotid sinus
- common carotid artery
- external jugular veins
- internal jugular vein
- brachiocephalic vein
- vertebral vein

_____

_____

_____

_____

_____

_____

_____

_____

_____

_____

• = _Term boldfaced in this module_

17.17

The internal carotid arteries and the vertebral arteries supply the brain, which is drained by the dural sinuses and the internal jugular veins

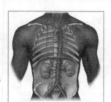

- ophthalmic artery
- anterior cerebral artery
- middle cerebral artery
- cerebral arterial circle
- great cerebral vein

- straight sinus
- cavernous sinus
- superior sagittal sinus
- vertebral vein

- petrosal sinuses
- transverse sinuses
- sigmoid sinuses

17.18

The regions supplied by the descending aorta are drained by the superior and inferior venae cavae

- thoracic aorta
- abdominal aorta
- intercostal arteries
- superior phrenic arteries
- bronchial arteries
- esophageal arteries
- mediastinal arteries
- pericardial arteries
- inferior phrenic arteries
- adrenal arteries
- renal arteries

- gonadal arteries
- lumbar arteries
- celiac trunk
- superior mesenteric artery
- inferior mesenteric artery
- azygos vein
- hemiazygos vein
- intercostal veins
- esophageal veins

- bronchial veins
- mediastinal veins
- lumbar veins
- gonadal veins
- hepatic veins
- renal veins
- adrenal veins
- phrenic veins

17.19

The viscera supplied by the celiac trunk and mesenteric arteries are drained by the tributaries of the hepatic portal vein

- celiac trunk
- common hepatic artery
- left gastric artery
- splenic artery
- superior mesenteric artery
- inferior mesenteric artery

- hepatic portal vein
- superior mesenteric vein
- inferior mesenteric vein

- splenic vein
- gastric veins
- cystic vein

• = *Term boldfaced in this module*

17.20

The pelvis and lower limbs are supplied by branches of the common iliac arteries and drained by tributaries of the common iliac veins

- right common iliac artery
- left common iliac artery
- internal iliac artery
- external iliac artery
- femoral artery
- deep femoral artery
- femoral circumflex arteries
- popliteal artery
- posterior tibial artery
- anterior tibial artery
- dorsalis pedis

- dorsal arch
- plantar arch
- fibular artery
- common iliac vein
- plantar venous arch
- anterior tibial vein
- posterior tibial vein
- fibular vein
- dorsal venous arch
- great saphenous vein

- small saphenous vein
- deep femoral vein
- femoral circumflex vein
- external iliac vein
- popliteal vein
- femoral vein

17.21

The pattern of blood flow through the fetal heart and the systemic circuit must change at birth

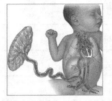

- umbilical arteries
- umbilical vein
- ductus venosus
- foramen ovale
- ductus arteriosus
- fossa ovalis
- ligamentum arteriosum
- ventricular septal defects

- patent
- patent foramen ovale
- patent ductus arteriosus
- tetralogy of Fallot

- atrioventricular septal defect
- transposition of great vessels

• = *Term boldfaced in this module*

1. Labeling

Label each of the structures in the accompanying figure.

a
b
c
d
e
f
g
h
i
j
k
l
m

n
o
p
q
r
s
t
u
v
w
x

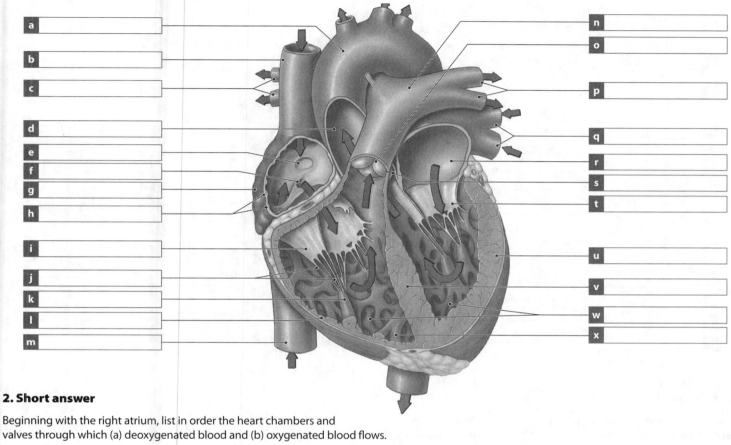

2. Short answer

Beginning with the right atrium, list in order the heart chambers and
valves through which (a) deoxygenated blood and (b) oxygenated blood flows.

3. Matching

Match the following terms with the most closely related description.

- cardiac skeleton
- fossa ovalis
- intercalated discs
- serous membrane
- tricuspid valve
- aortic valve
- endocardium
- aorta
- myocardium
- anastomoses
- coronary sinus
- calcium ions

a _____ Blood to systemic arteries
b _____ Activates contraction
c _____ Muscular wall of heart
d _____ Returns blood from heart back to heart
e _____ Depression in interatrial septum
f _____ Right atrioventricular valve
g _____ Interconnections betwen blood vessels
h _____ Cardiac muscle fiber connections
i _____ Pericardium
j _____ Inner surface of heart
k _____ Stabilizing connective tissue
l _____ Semilunar valve

1. Labeling

Label each of the structures in the accompanying figure.

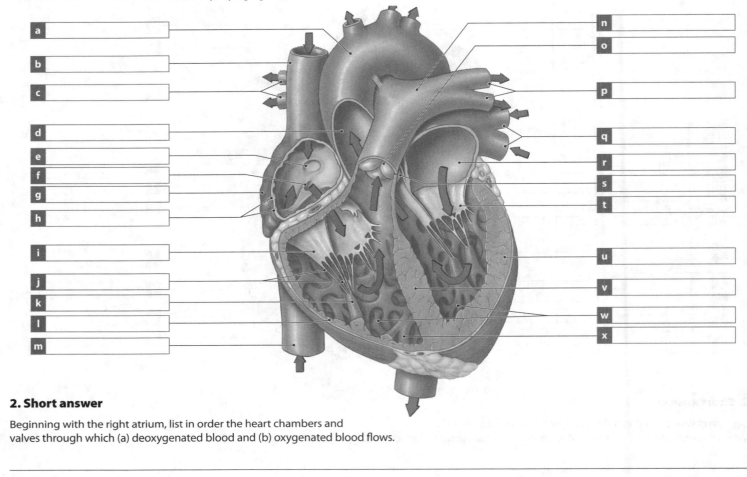

a _____
b _____
c _____
d _____
e _____
f _____
g _____
h _____
i _____
j _____
k _____
l _____
m _____

n _____
o _____
p _____
q _____
r _____
s _____
t _____
u _____
v _____
w _____
x _____

2. Short answer

Beginning with the right atrium, list in order the heart chambers and
valves through which (a) deoxygenated blood and (b) oxygenated blood flows.

_____

_____

_____

3. Matching

Match the following terms with the most closely related description.

- cardiac skeleton
- fossa ovalis
- intercalated discs
- serous membrane
- tricuspid valve
- aortic valve
- endocardium
- aorta
- myocardium
- anastomoses
- coronary sinus
- calcium ions

a _____   Blood to systemic arteries

b _____   Activates contraction

c _____   Muscular wall of heart

d _____   Returns blood from heart back to heart

e _____   Depression in interatrial septum

f _____   Right atrioventricular valve

g _____   Interconnections betwen blood vessels

h _____   Cardiac muscle fiber connections

i _____   Pericardium

j _____   Inner surface of heart

k _____   Stabilizing connective tissue

l _____   Semilunar valve

1. Short answer

Refer to the accompanying graph of the cardiac cycle in answering questions a through f below.

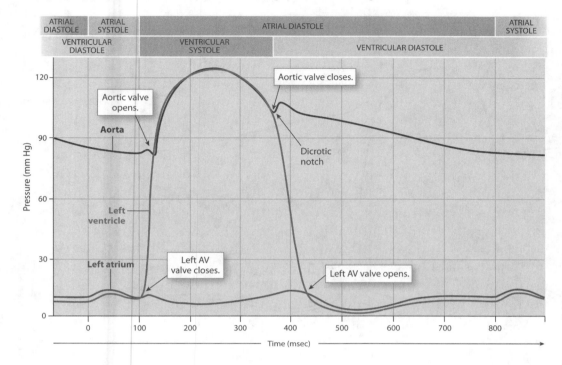

a What event occurs when the pressure in the left ventricle rises above that in the left atrium? _____

b During ventricular systole, the blood volume in the atria is _____ (increasing/decreasing), and the volume in the ventricle is _____ (increasing/decreasing).

c During most of ventricular diastole, the pressure in the left ventricle is _____ (greater than/the same as/less than) the pressure in the left atrium.

d What event occurs when the pressure within the left ventricle becomes greater than the pressure within the aorta? _____

e During isovolumetric contraction, pressure is highest in the _____ .

f During what part of the cardiac cycle is blood pressure highest in the large systemic arteries? _____

2. Matching

Match the following terms with the most closely related description.

- P wave
- cardiac output
- automaticity
- "lubb" sound
- "dubb" sound
- sympathetic neurons
- stroke volume
- tachycardia
- bradycardia
- parasympathetic neurons

a _____ AV valves close

b _____ Self-stimulated cardiac muscle contractions

c _____ Atrial depolarization

d _____ Amount of blood ejected by ventricle during a single beat

e _____ HR x SV

f _____ Decrease the heart rate

g _____ Term for slower-than-normal HR

h _____ Increase the heart rate

i _____ Semilunar valve closes

j _____ Term for faster-than-normal HR

1. Short answer

Refer to the accompanying graph of the cardiac cycle in answering questions a through f below.

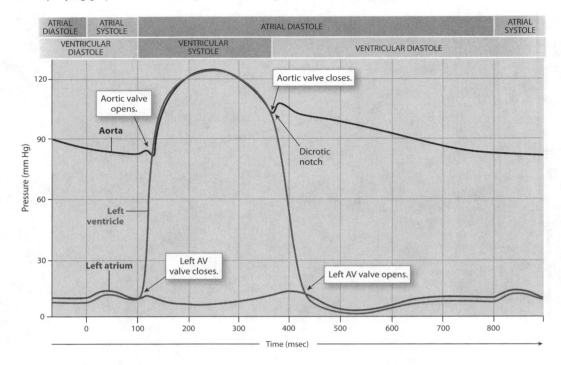

a What event occurs when the pressure in the left ventricle rises above that in the left atrium? _____

b During ventricular systole, the blood volume in the atria is _____ (increasing/decreasing), and the volume in the ventricle is _____ (increasing/decreasing).

c During most of ventricular diastole, the pressure in the left ventricle is _____ (greater than/the same as/less than) the pressure in the left atrium.

d What event occurs when the pressure within the left ventricle becomes greater than the pressure within the aorta? _____

e During isovolumetric contraction, pressure is highest in the _____ .

f During what part of the cardiac cycle is blood pressure highest in the large systemic arteries? _____

2. Matching

Match the following terms with the most closely related description.

- P wave
- cardiac output
- automaticity
- "lubb" sound
- "dubb" sound
- sympathetic neurons
- stroke volume
- tachycardia
- bradycardia
- parasympathetic neurons

a _____ AV valves close

b _____ Self-stimulated cardiac muscle contractions

c _____ Atrial depolarization

d _____ Amount of blood ejected by ventricle during a single beat

e _____ HR x SV

f _____ Decrease the heart rate

g _____ Term for slower-than-normal HR

h _____ Increase the heart rate

i _____ Semilunar valve closes

j _____ Term for faster-than-normal HR

1. Short answer

What three primary factors influence blood pressure and blood flow? _____

2. Matching

Match the following terms with the most closely related description.

- autoregulation
- venous return
- local vasodilators
- natriuretic peptides
- chemoreceptors
- baroreceptors
- turbulence
- net hydrostatic pressure
- medulla oblongata
- edema
- viscosity
- osmotic pressure

a	_____	Detect changes in pressure
b	_____	Vasomotor center
c	_____	Aided by thoracic pressure changes due to breathing and skeletal muscle activity
d	_____	Causes immediate, local homeostatic responses
e	_____	Decreased tissue O_2 and increased CO_2
f	_____	Carotid bodies
g	_____	Forces water into a capillary
h	_____	Opposite of smooth blood flow
i	_____	Resistance to flow
j	_____	Forces water out of a capillary
k	_____	Peripheral vasodilation
l	_____	Excess interstitial fluid accumulation

3. Multiple choice

Choose the bulleted item that best completes each statement.

a Of the following blood vessels, the greatest drop in blood pressure occurs in the _____ .

- capillaries
- veins
- venules
- arterioles

b The central regulation of cardiac output primarily involves the _____ .

- somatic nervous system
- central nervous system
- autonomic nervous system
- all of these

c Hormonal regulation by ADH, epinephrine, angiotensin II, and norepinephrine results in _____ .

- decreasing peripheral vasoconstriction
- increasing peripheral vasoconstriction
- increasing peripheral vasodilation
- all of these

d The three primary interrelated changes that occur as exercise begins are _____ .

- decreased vasodilation, increased venous return, increased cardiac output
- increased vasodilation, decreased venous return, increased cardiac output
- decreased vasodilation, decreased venous return, decreased cardiac output
- increased vasodilation, increased venous return, increased cardiac output

e The only part of the body where the blood supply is unchanged during exercise at maximal levels is the _____ .

- heart
- brain
- kidney
- skin

f The systems responsible for modifying heart rate and regulating blood pressure are the _____ systems.

- respiratory and muscular
- respiratory and nervous
- muscular and urinary
- nervous and endocrine

1. Short answer

What three primary factors influence blood pressure and blood flow? _____

2. Matching

Match the following terms with the most closely related description.

- autoregulation
- venous return
- local vasodilators
- natriuretic peptides
- chemoreceptors
- baroreceptors
- turbulence
- net hydrostatic pressure
- medulla oblongata
- edema
- viscosity
- osmotic pressure

a	_____	Detect changes in pressure
b	_____	Vasomotor center
c	_____	Aided by thoracic pressure changes due to breathing and skeletal muscle activity
d	_____	Causes immediate, local homeostatic responses
e	_____	Decreased tissue O_2 and increased CO_2
f	_____	Carotid bodies
g	_____	Forces water into a capillary
h	_____	Opposite of smooth blood flow
i	_____	Resistance to flow
j	_____	Forces water out of a capillary
k	_____	Peripheral vasodilation
l	_____	Excess interstitial fluid accumulation

3. Multiple choice

Choose the bulleted item that best completes each statement.

a Of the following blood vessels, the greatest drop in blood pressure occurs in the _____ .

- capillaries
- veins
- venules
- arterioles

b The central regulation of cardiac output primarily involves the _____ .

- somatic nervous system
- central nervous system
- autonomic nervous system
- all of these

c Hormonal regulation by ADH, epinephrine, angiotensin II, and norepinephrine results in _____ .

- decreasing peripheral vasoconstriction
- increasing peripheral vasoconstriction
- increasing peripheral vasodilation
- all of these

d The three primary interrelated changes that occur as exercise begins are _____ .

- decreased vasodilation, increased venous return, increased cardiac output
- increased vasodilation, decreased venous return, increased cardiac output
- decreased vasodilation, decreased venous return, decreased cardiac output
- increased vasodilation, increased venous return, increased cardiac output

e The only part of the body where the blood supply is unchanged during exercise at maximal levels is the _____ .

- heart
- brain
- kidney
- skin

f The systems responsible for modifying heart rate and regulating blood pressure are the _____ systems.

- respiratory and muscular
- respiratory and nervous
- muscular and urinary
- nervous and endocrine

Visual Outline with Key Terms

Summarize the content of each module using the terms in the order provided.

The Structure of the Heart

- base
- apex
- superior border
- right border
- left border
- inferior border

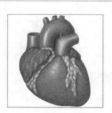

_____

_____

_____

_____

_____

_____

_____

_____

18.1

The wall of the heart contains concentric layers of cardiac muscle tissue

- myocardium
- endocardium
- parietal pericardium
- pericardial sac
- epicardium
- intercalated disc
- functional syncytium

_____

_____

_____

_____

_____

_____

_____

18.2

The heart is located in the mediastinum, suspended within the pericardial cavity

- mediastinum
- great vessels
- pericarditis
- cardiac tamponade

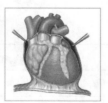

_____

_____

_____

_____

_____

_____

_____

• = _Term boldfaced in this module_

18.3

The boundaries between the chambers of the heart can be distinguished on its external surface

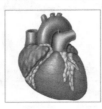

- anterior surface
- sulci
- auricle
- coronary sulcus
- ligamentum arteriosum
- anterior interventricular sulcus
- posterior surface
- coronary sinus
- posterior interventricular sulcus

_____

_____

_____

_____

_____

_____

_____

_____

_____

18.4

The heart has an extensive blood supply

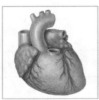

- coronary arteries
- right coronary artery
- marginal arteries
- left coronary artery
- anterior interventricular artery
- circumflex artery
- posterior interventricular artery
- anterior cardiac veins
- great cardiac vein
- coronary sinus
- posterior cardiac vein
- small cardiac vein
- middle cardiac vein
- elastic rebound

_____

_____

_____

_____

_____

_____

_____

_____

_____

_____

_____

18.5

Internal valves control the direction of blood flow between the heart chambers

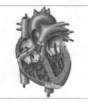

- interatrial septum
- interventricular septum
- atrioventricular (AV) valves
- fossa ovalis
- pectinate muscles
- right atrioventricular (AV) valve
- tricuspid valve
- cusps
- chordae tendineae
- papillary muscles
- pulmonary valve
- moderator band
- left atrioventricular (AV) valve
- bicuspid valve
- mitral valve
- trabeculae carneae
- aortic valve

_____

_____

_____

_____

_____

_____

_____

_____

_____

● = _Term boldfaced in this module_

18.6

When the heart beats, the AV valves close before the semilunar valves open, and the semilunar valves close before the AV valves open

- regurgitation
- aortic sinuses
- cardiac skeleton
- semilunar valves
- valvular heart disease (VHD)
- carditis

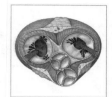

_____
_____
_____
_____
_____
_____
_____
_____

18.7

Arteriosclerosis can lead to coronary artery disease

- arteriosclerosis
- atherosclerosis
- plaque
- balloon angioplasty
- coronary artery disease (CAD)
- coronary ischemia
- thrombus
- digital subtraction angiography (DSA)
- stent

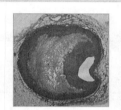

_____
_____
_____
_____
_____
_____
_____
_____

SECTION 2

The Cardiac Cycle

- cardiac cycle
- systole
- diastole

_____
_____
_____
_____
_____
_____

● = _Term boldfaced in this module_

18.8

The cardiac cycle creates pressure gradients that maintain blood flow

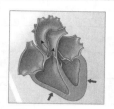

- atrial systole
- atrial diastole
- ventricular systole—first phase
- isovolumetric contraction
- ventricular systole—second phase
- ventricular ejection
- ventricular diastole—early
- isovolumetric relaxation
- ventricular diastole—late
- dicrotic notch
- heart sounds (S_1–S_4)

18.9

The heart rate, a key factor in cardiac output, is established by the SA node and distributed by the conducting system

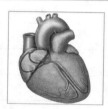

- cardiac output (CO)
- stroke volume
- automaticity
- conducting system
- sinoatrial node (SA node)
- internodal pathways
- atrioventricular node (AV node)
- AV bundle
- bundle branches
- Purkinje fibers

18.10

Cardiac muscle cell contractions last longer than skeletal muscle fiber contractions primarily due to differences in membrane permeability

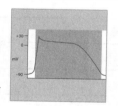

- fast sodium channels
- slow calcium channels
- absolute refractory period
- relative refractory period

• = *Term boldfaced in this module*

18.11

The intrinsic heart rate can be altered by autonomic activity

- prepotential
- pacemaker potential
- bradycardia
- tachycardia
- cardioinhibitory center
- cardioacceleratory center

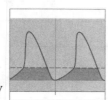

18.12

Stroke volume depends on the relationship between EDV and ESV

- stroke volume
- end-diastolic volume (EDV)
- end-systolic volume (ESV)
- venous return
- filling time
- preload
- contractility
- afterload

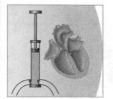

18.13

Cardiac output is regulated by adjustments in heart rate and stroke volume

- heart failure

• = *Term boldfaced in this module*

18.14

Normal and abnormal cardiac activity can be detected in an electrocardiogram

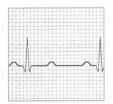

- electrocardiogram (ECG or EKG)
- P wave
- QRS complex
- R wave
- T wave
- P–R interval
- Q–T interval
- cardiac arrhythmias
- premature atrial contractions (PACs)
- paroxysmal atrial tachycardia (PAT)
- atrial fibrillation (AF)
- premature ventricular contractions (PVCs)
- ectopic pacemaker
- ventricular tachycardia (VT or V-tach)
- ventricular fibrillation (VF)
- cardiac arrest

SECTION 3

The Coordination of Cardiac Output and Peripheral Blood Flow

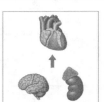

- cardiac output
- blood pressure
- arterial pressure
- venous pressure
- resistance
- peripheral resistance
- capillary exchange
- venous return
- neural and hormonal regulation

18.15

There are multiple sources of resistance within the cardiovascular system, but vessel diameter is the primary factor under normal conditions

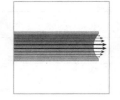

- total peripheral resistance
- vascular resistance
- viscosity
- turbulence

• = *Term boldfaced in this module*

18.16

Blood flow is determined by the interplay between arterial pressure and peripheral resistance

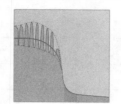

- blood flow (F)
- blood pressure
- peripheral resistance
- pressure gradient
- systolic pressure
- diastolic pressure
- pulse pressure
- mean arterial pressure (MAP)
- capillary exchange
- capillary hydrostatic pressure (CHP)
- filtration

18.17

Capillary exchange is a dynamic process that includes diffusion, filtration, and reabsorption

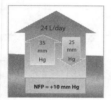

- capillary hydrostatic pressure (CHP)
- blood colloid osmotic pressure (BCOP)
- net filtration pressure (NFP)
- recall of fluids
- edema

18.18

Cardiovascular regulatory mechanisms respond to changes in blood pressure or blood composition

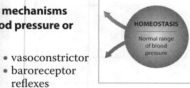

- tissue perfusion
- autoregulation
- vasodilators
- central regulation
- cardioacceleratory centers
- vasoconstrictor
- baroreceptor reflexes

• = *Term boldfaced in this module*

18.19

The endocrine responses to low blood pressure and low blood volume are very different from those to high blood pressure and high blood volume

- angiotensin converting enzyme (ACE)
- natriuretic peptides
- atrial natriuretic peptide (ANP)
- brain natriuretic peptide (BNP)

18.20

Chemoreceptors monitor the chemical composition of the blood and CSF

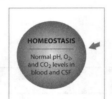

- chemoreceptor reflexes
- carotid bodies
- aortic bodies

18.21

The cardiovascular centers make extensive adjustments to cardiac output and blood distribution during exercise

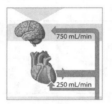

- respiratory pump

18.22

Short-term and long-term mechanisms compensate for a reduction in blood volume

- shock
- hypotension
- circulatory shock
- progressive shock
- irreversible shock
- circulatory collapse

● = *Term boldfaced in this module*

1. Labeling

Label the structures of the lymphatic system in the accompanying figure.

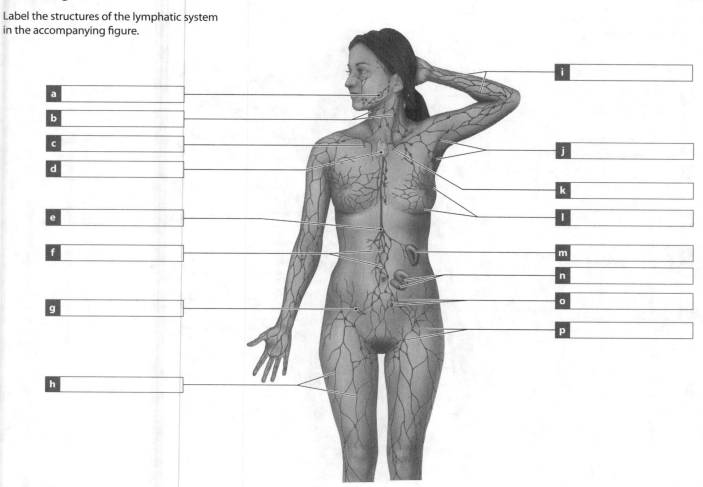

a	
b	
c	
d	
e	
f	
g	
h	
i	
j	
k	
l	
m	
n	
o	
p	

2. Matching

Match the following terms with the most closely related description.

- B cells
- spleen
- lymphatic capillaries
- cytotoxic T cells
- reticular epithelial cells
- tonsils
- lymphopoiesis
- thymic corpuscles
- lymphoid organs
- helper T cells and suppressor T cells
- afferent lymphatics
- right subclavian vein
- thoracic duct
- lymph nodes

a _____ Beginning of lymphatic system

b _____ Thymus medullary cells

c _____ Receives lymph from right lymphatic duct

d _____ Receives lymph from left half of head

e _____ Thymus, spleen, and lymph nodes

f _____ Smallest lymphoid organs

g _____ Regulate and coordinate the immune response

h _____ Largest mass of lymphoid tissue in body

i _____ Maintains blood–thymus barrier

j _____ Occurs in red bone marrow, thymus, and lymphoid tissues

k _____ Cell-mediated immunity

l _____ Lymphoid nodules in walls of pharynx

m _____ Antibody-mediated immunity

n _____ Carry lymph to lymph nodes

1. Labeling

Label the structures of the lymphatic system in the accompanying figure.

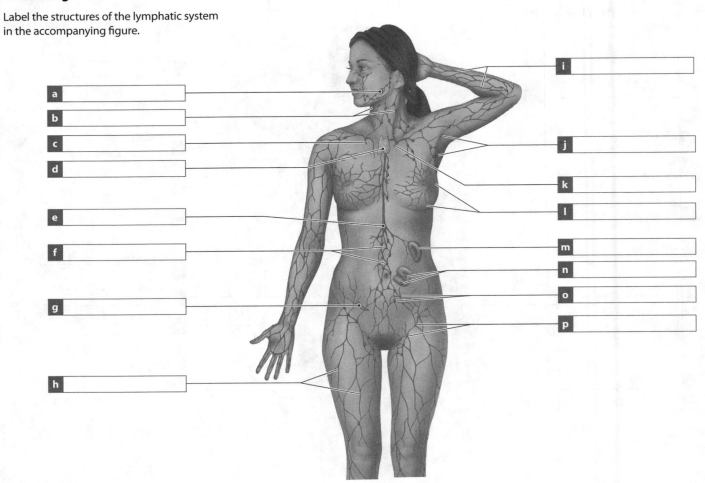

a

b

c

d

e

f

g

h

i

j

k

l

m

n

o

p

2. Matching

Match the following terms with the most closely related description.

- B cells
- spleen
- lymphatic capillaries
- cytotoxic T cells
- reticular epithelial cells
- tonsils
- lymphopoiesis
- thymic corpuscles
- lymphoid organs
- helper T cells and suppressor T cells
- afferent lymphatics
- right subclavian vein
- thoracic duct
- lymph nodes

a _____ Beginning of lymphatic system

b _____ Thymus medullary cells

c _____ Receives lymph from right lymphatic duct

d _____ Receives lymph from left half of head

e _____ Thymus, spleen, and lymph nodes

f _____ Smallest lymphoid organs

g _____ Regulate and coordinate the immune response

h _____ Largest mass of lymphoid tissue in body

i _____ Maintains blood–thymus barrier

j _____ Occurs in red bone marrow, thymus, and lymphoid tissues

k _____ Cell-mediated immunity

l _____ Lymphoid nodules in walls of pharynx

m _____ Antibody-mediated immunity

n _____ Carry lymph to lymph nodes

1. Labeling

Fill in the spaces with the name of the nonspecific defense described at right.

a _____ _____

b _____

c _____ _____

d _____

e _____

f _____ _____

g _____

a keep hazardous organisms and materials outside the body. For example, a mosquito that lands on your head may be unable to reach the surface of the scalp if you have a full head of hair.

b are cells that engulf pathogens and cell debris. Examples are the macrophages of peripheral tissues and the eosinophils and neutrophils of blood.

c is the destruction of abnormal cells by NK cells in peripheral tissues.

Destruction of abnormal cells

d are chemical messengers that coordinate the defenses against viral infections.

e is a system of circulating proteins that assists antibodies in the destruction of pathogens.

f is a localized, tissue-level response that tends to limit the spread of an injury or infection.

Inflammation

g is an elevation of body temperature that accelerates tissue metabolism and the activity of defenses.

2. Multiple choice

Choose the bulleted item that best completes each statement.

a A physical barrier such as the skin provides a nonspecific body defense due to its makeup, which includes _____.

- multiple layers
- a coating of keratinized cells
- a network of desmosomes locking adjacent cells together
- all of these

b NK (natural killer) cells sensitive to the presence of abnormal plasma membranes are primarily involved in _____.

- defenses against specific threats
- phagocytic activity for defense
- complex, time-consuming defense mechanisms
- immunological surveillance

c The nonspecific defense that breaks down cells, attracts phagocytes, and stimulates inflammation is _____.

- the inflammatory response
- the action of interferons
- the complement system
- immunological surveillance

d The protein(s) that interfere with the replication of viruses is (are) _____.

- complement proteins
- heparin
- pyrogens
- interferons

e Circulating proteins that reset the thermostat in the hypothalamus, causing a rise in body temperature, are called _____.

- pyrogens
- interferons
- lysosomes
- complement proteins

f The "first line of cellular defense" against pathogenic invasion is _____.

- phagocytes
- mucus
- hair
- interferon

3. Short answer

We usually associate a fever with illness or disease. How may a fever be beneficial?

_____

1. Labeling

Fill in the spaces with the name of the nonspecific defense described at right.

a _____ _____

b _____

c _____ _____

d _____

e _____

f _____ _____

g _____

a keep hazardous organisms and materials outside the body. For example, a mosquito that lands on your head may be unable to reach the surface of the scalp if you have a full head of hair.

b are cells that engulf pathogens and cell debris. Examples are the macrophages of peripheral tissues and the eosinophils and neutrophils of blood.

c is the destruction of abnormal cells by NK cells in peripheral tissues.

→ Destruction of abnormal cells

d are chemical messengers that coordinate the defenses against viral infections.

e is a system of circulating proteins that assists antibodies in the destruction of pathogens.

f is a localized, tissue-level response that tends to limit the spread of an injury or infection.

→ Inflammation

g is an elevation of body temperature that accelerates tissue metabolism and the activity of defenses.

2. Multiple choice

Choose the bulleted item that best completes each statement.

a A physical barrier such as the skin provides a nonspecific body defense due to its makeup, which includes _____.

- multiple layers
- a coating of keratinized cells
- a network of desmosomes locking adjacent cells together
- all of these

b NK (natural killer) cells sensitive to the presence of abnormal plasma membranes are primarily involved in _____.

- defenses against specific threats
- phagocytic activity for defense
- complex, time-consuming defense mechanisms
- immunological surveillance

c The nonspecific defense that breaks down cells, attracts phagocytes, and stimulates inflammation is _____.

- the inflammatory response
- the action of interferons
- the complement system
- immunological surveillance

d The protein(s) that interfere with the replication of viruses is (are) _____.

- complement proteins
- heparin
- pyrogens
- interferons

e Circulating proteins that reset the thermostat in the hypothalamus causing a rise in body temperature, are called _____

- pyrogens
- interferons
- lysosomes
- complement proteins

f The "first line of cellular defense" against pathogenic invasion is _____.

- phagocytes
- mucus
- hair
- interferon

3. Short answer

We usually associate a fever with illness or disease. How may a fever be beneficial?

_____

1. Matching

Match the following terms with the most closely related description.

- opsonization
- helper T cells
- antibody
- Class II MHC
- costimulation
- IgM
- Class I MHC
- IgG
- passive immunity
- anaphylaxis
- CD4 markers
- acquired immunity
- B lymphocytes

a _____ Two parallel pairs of polypeptide chains

b _____ Found on helper T cells

c _____ Active and passive

d _____ Transfer of antibodies

e _____ Attacked by HIV

f _____ Enhances phagocytosis

g _____ MHC proteins present in the plasma membranes of all nucleated cells

h _____ Differentiate into memory and plasma cells

i _____ MHC proteins present in the plasma membranes of all APCs and lymphocytes

j _____ Antibodies used to determine blood type

k _____ Secondary binding process required for T cell activation

l _____ Accounts for 80 percent of all immunoglobulins

m _____ Circulating allergen stimulates mast cells throughout body

2. Matching

Match the following terms with the most closely related description.

- cytotoxic T cells
- viruses
- B cells
- antibodies
- helper T cells
- macrophages
- natural killer (NK) cells
- suppressor T cells
- memory T cells and B cells

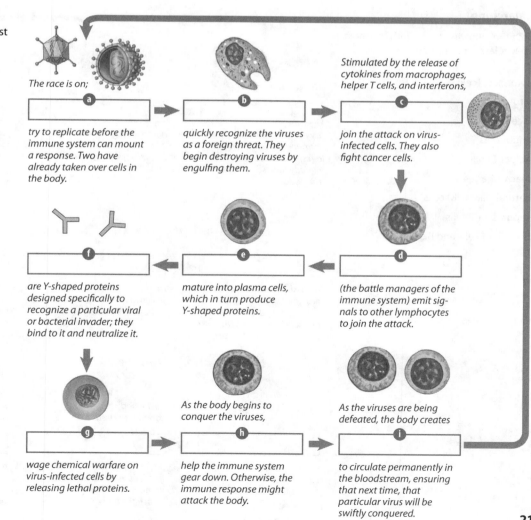

a _____ *The race is on;* *try to replicate before the immune system can mount a response. Two have already taken over cells in the body.*

b _____ *quickly recognize the viruses as a foreign threat. They begin destroying viruses by engulfing them.*

c _____ *Stimulated by the release of cytokines from macrophages, helper T cells, and interferons,* *join the attack on virus-infected cells. They also fight cancer cells.*

d _____ *(the battle managers of the immune system) emit signals to other lymphocytes to join the attack.*

e _____ *mature into plasma cells, which in turn produce Y-shaped proteins.*

f _____ *are Y-shaped proteins designed specifically to recognize a particular viral or bacterial invader; they bind to it and neutralize it.*

g _____ *wage chemical warfare on virus-infected cells by releasing lethal proteins.*

h _____ *As the body begins to conquer the viruses,* *help the immune system gear down. Otherwise, the immune response might attack the body.*

i _____ *As the viruses are being defeated, the body creates* *to circulate permanently in the bloodstream, ensuring that next time, that particular virus will be swiftly conquered.*

1. Matching

Match the following terms with the most closely related description.

- opsonization
- helper T cells
- antibody
- Class II MHC
- costimulation
- IgM
- Class I MHC
- IgG
- passive immunity
- anaphylaxis
- CD4 markers
- acquired immunity
- B lymphocytes

a _____ Two parallel pairs of polypeptide chains

b _____ Found on helper T cells

c _____ Active and passive

d _____ Transfer of antibodies

e _____ Attacked by HIV

f _____ Enhances phagocytosis

g _____ MHC proteins present in the plasma membranes of all nucleated cells

h _____ Differentiate into memory and plasma cells

i _____ MHC proteins present in the plasma membranes of all APCs and lymphocytes

j _____ Antibodies used to determine blood type

k _____ Secondary binding process required for T cell activation

l _____ Accounts for 80 percent of all immunoglobulins

m _____ Circulating allergen stimulates mast cells throughout body

2. Matching

Match the following terms with the most closely related description.

- cytotoxic T cells
- viruses
- B cells
- antibodies
- helper T cells
- macrophages
- natural killer (NK) cells
- suppressor T cells
- memory T cells and B cells

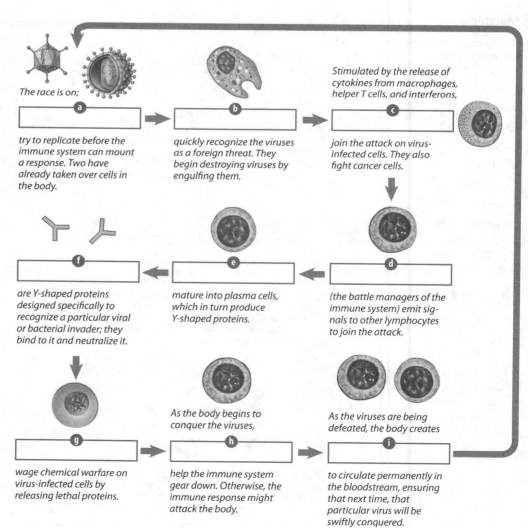

a _____
The race is on; ... try to replicate before the immune system can mount a response. Two have already taken over cells in the body.

b _____
quickly recognize the viruses as a foreign threat. They begin destroying viruses by engulfing them.

c _____
Stimulated by the release of cytokines from macrophages, helper T cells, and interferons, ... join the attack on virus-infected cells. They also fight cancer cells.

d _____
(the battle managers of the immune system) emit signals to other lymphocytes to join the attack.

e _____
mature into plasma cells, which in turn produce Y-shaped proteins.

f _____
are Y-shaped proteins designed specifically to recognize a particular viral or bacterial invader; they bind to it and neutralize it.

g _____
wage chemical warfare on virus-infected cells by releasing lethal proteins.

h _____
As the body begins to conquer the viruses, ... help the immune system gear down. Otherwise, the immune response might attack the body.

i _____
As the viruses are being defeated, the body creates ... to circulate permanently in the bloodstream, ensuring that next time, that particular virus will be swiftly conquered.

Visual Outline with Key Terms

Summarize the content of each module using the terms in the order provided.

Anatomy of the Lymphatic System

- ○ lymphatic system
- • lymphocytes
- • lymph
- • lymphatics

_____

19.1

Interstitial fluid continuously flows into lymphatic capillaries, and leaves tissue as lymph within lymphatic vessels

- • lymphatic vessels
- • lymphatic capillaries

19.2

Small lymphatic vessels collect to form lymphatic ducts that empty into the subclavian veins

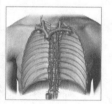

- • superficial lymphatics
- • deep lymphatics
- • lymphatic trunks
- • thoracic duct
- • right lymphatic duct
- • right and left jugular trunks

- • right and left subclavian trunks
- • right and left broncho-mediastinal trunks
- ○ right and left subclavian veins

- • cisterna chyli
- • lumbar trunks
- • intestinal trunk
- • lymphedema

• = *Term boldfaced in this module*

19.3

Lymphocytes are responsible for the immune functions of the lymphatic system

- antigens
- T cells
- cytotoxic T cells
- cell-mediated immunity
- helper T cells
- suppressor T cells
- B cells
- plasma cells
- antibody-mediated immunity
- NK (natural killer) cells
- immunological surveillance
- lymphopoiesis
- ○ red bone marrow
- lymphoid stem cells
- ○ thymus
- blood–thymus barrier

19.4

Lymphocytes aggregate within lymphoid tissues and lymphoid organs

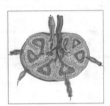

- lymphoid tissues
- lymphoid nodule
- lymphoid organs
- aggregated lymphoid nodules
- germinal center
- mucosa-associated lymphoid tissue (MALT)
- tonsils
- palatine tonsils
- pharyngeal tonsil
- lingual tonsils
- tonsillitis
- lymph nodes
- lymph glands
- afferent lymphatics
- dendritic cells
- outer cortex
- deep cortex
- medullary sinus
- efferent lymphatics
- hilum
- trabeculae
- appendicitis

19.5

The thymus is a lymphoid organ that produces functional T cells

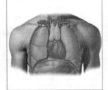

- ○ thymus
- thymosins
- involution
- lobes
- septa
- lobules
- cortex
- medulla
- reticular epithelial cells
- blood–thymus barrier
- thymic corpuscles

● = *Term boldfaced in this module*

19.6

The spleen, the largest lymphoid organ, responds to antigens in the bloodstream

- ○ spleen
- • gastrosplenic ligament
- • diaphragmatic surface
- • splenectomy
- • visceral surface
- • gastric area
- • renal area

- • hilum
- • pulp
- • red pulp
- • white pulp
- • trabeculae
- • trabecular arteries

- • central arteries
- • trabecular veins

_____

_____

Nonspecific Defenses

- ○ resistance
- • nonspecific defenses
- • nonspecific resistance
- • physical barriers
- • phagocytes
- • immunological surveillance

- • interferons
- • complement
- • inflammatory response
- • fever
- • specific defenses
- • immunity

- • specific resistance

_____

_____

19.7

Physical barriers prevent pathogens and toxins from entering body tissues and phagocytes provide the next line of defense

- ○ integumentary system
- ○ epithelia
- ○ phagocytes
- • neutrophils
- • eosinophils
- • macrophages

- • monocyte–macrophage system
- • fixed macrophages
- • free macrophages

- • emigration
- • diapedesis
- • chemotaxis

_____

_____

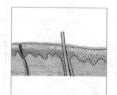

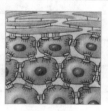

• = Term boldfaced in this module

19.8

NK cells perform immunological surveillance, detecting and destroying abnormal cells

- immunological surveillance
- tumor-specific antigens
- perforins
- immunological escape

19.9

Interferons and the complement system are distributed widely in body fluids

- interferons
- cytokines
- alpha (α)-interferons
- beta (β)-interferons
- gamma (γ)-interferons
- complement system
- classical pathway
- alternative pathway
- properdin
- opsonization

19.10

Inflammation is a localized tissue response to injury; fever is a generalized response to tissue damage and infection

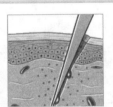

- inflammation
- inflammatory response
- fever
- pyrogens

• = *Term boldfaced in this module*

Specific Defenses

- specific defenses
- immunity
- acquired immunity
- passive immunity
- naturally acquired passive immunity
- artificially acquired passive immunity
- active immunity (immune response)
- naturally acquired active immunity
- artificially acquired active immunity
- innate immunity
- specificity
- versatility
- clone
- immunologic memory

Specific Defenses (Immunity)

Respond to threats on an individualized basis

- memory cells
- tolerance

19.11

Specific defenses are triggered by exposure to antigenic fragments

- antigen presentation
- major histocompatibility complex (MHC)
- MHC proteins
- Class I MHC proteins
- Class II MHC proteins
- antigen-presenting cells (APCs)

19.12

Infected cells stimulate the formation and division of cytotoxic T cells, memory T$_C$ cells, and suppressor T cells

- antigen recognition
- CD markers
- CD8 markers
- CD8 T cells
- CD4 markers
- CD4 T cells
- costimulation
- cytotoxic T cells (T$_C$ cells)
- memory T$_C$ cells
- suppressor T cells (T$_S$ cells)
- suppression factors
- apoptosis
- lymphotoxin

• = *Term boldfaced in this module*

19.13

APCs can stimulate the activation of CD4 T cells; this produces helper T cells that promote B cell activation and antibody production

- helper T cells (T_H cells)
- memory T_H cells
- cytokines
- sensitization
- memory B cells
- plasma cells

19.14

Antibodies are small soluble proteins that bind to specific antigens; they may inactivate the antigens or trigger another defensive process

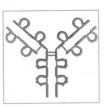

- antibody molecule
- heavy chains
- light chains
- constant segments
- variable segments
- antigen binding sites
- antigen-antibody complex
- antigenic determinant sites
- complete antigen
- partial antigens
- immunoglobulins (Igs)
- IgG, IgE, IgD, IgM, IgA
- primary response
- antibody titer
- secondary response

19.15

Antibodies use many different mechanisms to destroy target antigens

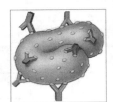

- neutralization
- opsonization
- immune complex
- agglutination

• = *Term boldfaced in this module*

19.16

Allergies and anaphylaxis are caused by antibody responses

- allergies
- allergens
- immediate hypersensitivity
- allergic rhinitis
- hypersensitivity reaction
- anaphylaxis
- anaphylactic shock
- localized allergic reactions

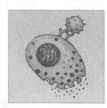

- systemic allergic reactions

_____

19.17

Specific and nonspecific defenses work together to defeat pathogens

- nonspecific defenses
- specific defenses

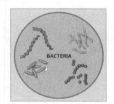

_____

19.18

Immune disorders involving both overactivity and underactivity can be harmful

- autoimmune disorders
- autoantibodies
- thyroiditis
- rheumatoid arthritis
- insulin-dependent diabetes mellitus (IDDM)
- graft rejection
- immunosuppression
- cyclosporin A (CsA)
- allergies
- immuno-deficiency diseases
- acquired immune deficiency syndrome (AIDS)

- human immunodeficiency virus (HIV)
- opportunistic infections

_____

• = _Term boldfaced in this module_

1. Labeling

Label each of the respiratory system structures in the following figure.

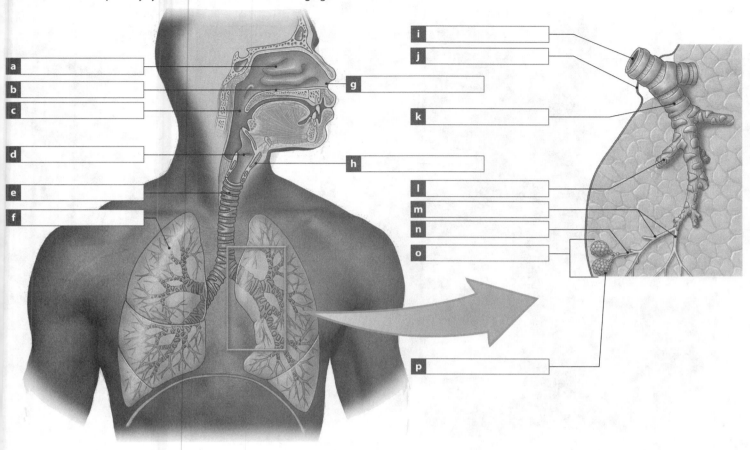

2. Matching

Match the following terms with the most closely related description.

- respiratory bronchiole
- respiratory mucosa
- phonation
- bronchodilation
- terminal bronchiole
- laryngeal prominence
- type I pneumocytes
- type II pneumocytes
- cystic fibrosis
- trachea
- pharynx
- respiratory membrane
- larynx
- bronchoconstriction

a	_____	Produce surfactant
b	_____	Windpipe
c	_____	Simple squamous epithelial cells
d	_____	Sympathetic activation
e	_____	Supplies a pulmonary lobule
f	_____	Parasympathetic activation
g	_____	Start of respiratory portion of respiratory tract
h	_____	Gas exchange
i	_____	Sound production at the larynx
j	_____	Chamber shared by respiratory and digestive systems
k	_____	Surrounds and protects the glottis
l	_____	Lethal inherited respiratory disease
m	_____	Lines conducting portion of respiratory tract
n	_____	Anterior surface of thyroid cartilage

1. Labeling

Label each of the respiratory system structures in the following figure.

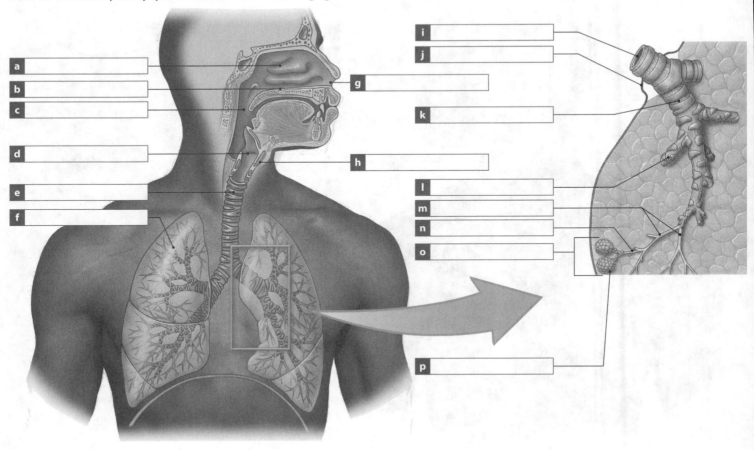

2. Matching

Match the following terms with the most closely related description.

- respiratory bronchiole
- respiratory mucosa
- phonation
- bronchodilation
- terminal bronchiole
- laryngeal prominence
- type I pneumocytes
- type II pneumocytes
- cystic fibrosis
- trachea
- pharynx
- respiratory membrane
- larynx
- bronchoconstriction

a _____ Produce surfactant

b _____ Windpipe

c _____ Simple squamous epithelial cells

d _____ Sympathetic activation

e _____ Supplies a pulmonary lobule

f _____ Parasympathetic activation

g _____ Start of respiratory portion of respiratory tract

h _____ Gas exchange

i _____ Sound production at the larynx

j _____ Chamber shared by respiratory and digestive systems

k _____ Surrounds and protects the glottis

l _____ Lethal inherited respiratory disease

m _____ Lines conducting portion of respiratory tract

n _____ Anterior surface of thyroid cartilage

1. Short answer

Identify and describe the various pulmonary volumes and capacities indicated in the following graph.

a _____

b _____

c _____

d _____

e _____

f _____

g _____

h _____

i _____

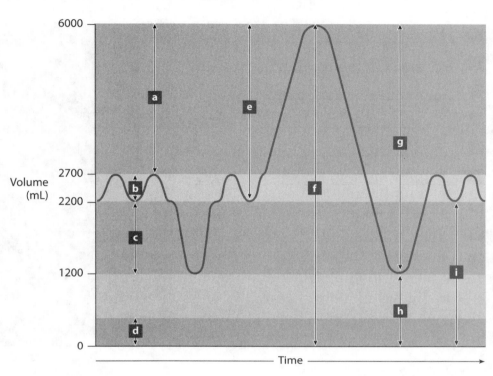

2. Matching

Match the following terms with the most closely related description.

- apnea
- hemoglobin releases more O_2
- lowers vital capacity
- external intercostals
- iron ion
- Boyle's law
- compliance
- bicarbonate ion
- anoxia
- atelectasis
- hypocapnia
- pneumotaxic centers
- apneustic centers
- partial pressure

a _____ Single gas in a mixture

b _____ A cause of tissue death

c _____ Inverse pressure/volume relationship

d _____ Expandability of lungs

e _____ CO_2 transport

f _____ Elastic tissue deterioration

g _____ Act to elevate ribs

h _____ Heme unit

i _____ Blood pH decreases

j _____ Promotes passive or active exhalation

k _____ Stimulates DRG and promotes inhalation

l _____ Hyperventilation

m _____ Collapsed lung

n _____ Period of suspended respiration

3. Section integration

Compare and contrast external respiration, pulmonary ventilation, and internal respiration. _____

_____

_____

1. Short answer

Identify and describe the various pulmonary volumes and capacities indicated in the following graph.

a _____

b _____

c _____

d _____

e _____

f _____

g _____

h _____

i _____

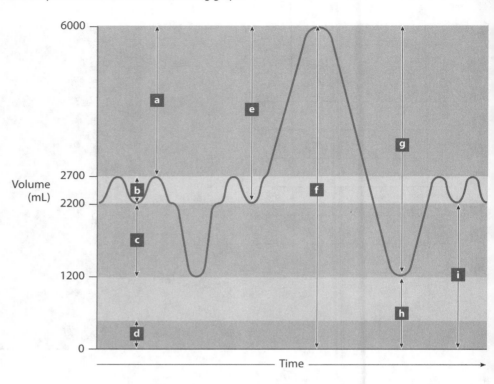

2. Matching

Match the following terms with the most closely related description.

- apnea
- hemoglobin releases more O_2
- lowers vital capacity
- external intercostals
- iron ion
- Boyle's law
- compliance
- bicarbonate ion
- anoxia
- atelectasis
- hypocapnia
- pneumotaxic centers
- apneustic centers
- partial pressure

a _____ Single gas in a mixture

b _____ A cause of tissue death

c _____ Inverse pressure/volume relationship

d _____ Expandability of lungs

e _____ CO_2 transport

f _____ Elastic tissue deterioration

g _____ Act to elevate ribs

h _____ Heme unit

i _____ Blood pH decreases

j _____ Promotes passive or active exhalation

k _____ Stimulates DRG and promotes inhalation

l _____ Hyperventilation

m _____ Collapsed lung

n _____ Period of suspended respiration

3. Section integration

Compare and contrast external respiration, pulmonary ventilation, and internal respiration. _____

_____

_____

Visual Outline with Key Terms

Summarize the content of each module using the terms in the order provided.

Functional Anatomy of the Respiratory System

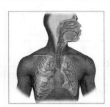

- respiratory system
- respiratory tract
- conducting portion
- bronchioles
- respiratory portion
- alveoli
- upper respiratory system
- lower respiratory system

_____

_____

_____

_____

_____

_____

_____

20.1

The respiratory mucosa is protected by the respiratory defense system

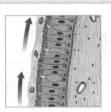

- respiratory defense system
- respiratory mucosa
- mucus escalator
- lamina propria
- cystic fibrosis (CF)

_____

_____

_____

_____

_____

_____

20.2

The upper portions of the respiratory system include the nose, nasal cavity, paranasal sinuses, and pharynx

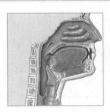

- bridge of the nose
- nasal cartilages
- external nares
- superior, middle, and inferior meatuses
- nasal septum
- ○ paranasal sinuses
- pharynx
- nasopharynx
- oropharynx
- laryngopharynx
- trachea
- internal nares
- nasal vestibule
- hard palate
- soft palate
- glottis
- larynx

_____

_____

_____

_____

_____

_____

_____

_____

• = _Term boldfaced in this module_

20.3

The larynx protects the glottis and produces sounds

- larynx
- epiglottis
- thyroid cartilage
- cricoid cartilage
- vestibular ligaments
- vocal ligaments
- cuneiform cartilages

- corniculate cartilages
- arytenoid cartilages
- vocal folds
- vocal ligaments
- vocal cords

- vestibular folds
- phonation
- articulation

_____
_____
_____
_____
_____
_____
_____
_____
_____

20.4

The trachea and primary bronchi convey air to and from the lungs

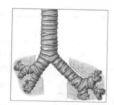

- trachea
- primary bronchi
- tracheal cartilages
- trachealis muscle
- secondary bronchi
- tertiary bronchi

- bronchioles
- terminal bronchioles
- pulmonary lobule
- root of the lung
- bronchodilation

- bronchoconstriction
- asthma

_____
_____
_____
_____
_____
_____
_____
_____

20.5

The lungs contain lobes that are subdivided into bronchopulmonary segments

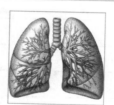

- bronchopulmonary segment
- bronchial tree
- right lung
- horizontal and oblique fissures

- superior, middle, and inferior lobes
- cardiac notch
- left lung
- oblique fissure

- superior and inferior lobes
- hilum

_____
_____
_____
_____
_____
_____
_____

• = *Term boldfaced in this module*

Pulmonary lobules contain alveoli, where gas exchange occurs

- alveoli
- pulmonary lobule
- respiratory bronchiole
- alveolar ducts
- alveolar sac
- visceral pleura
- pleural fluid
- parietal pleura
- type II pneumocytes
- surfactant
- type I pneumocytes
- respiratory membrane

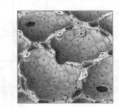

_____

_____

_____

_____

_____

_____

_____

_____

SECTION 2

Respiratory Physiology

- respiration
- external respiration
- pulmonary ventilation
- alveolar ventilation
- gas diffusion
- internal respiration
- hypoxia
- anoxia

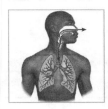

_____

_____

_____

_____

_____

_____

Pulmonary ventilation is driven by pressure changes within the pleural cavities

- Boyle's law
- atelectasis
- intrapulmonary pressure
- tidal volume (V_T)
- gas pressures
- millimeters of mercury (mm Hg)
- torr
- centimeters of water (cm H_2O)
- pounds per square inch (psi)

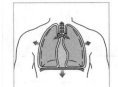

_____

_____

_____

_____

_____

_____

_____

_____

• = *Term boldfaced in this module*

20.8

Respiratory muscles in various combinations adjust the tidal volume to meet respiratory demands

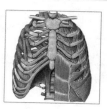

- inspiratory muscles
- expiratory muscles
- primary respiratory muscles
- accessory respiratory muscles
  - diaphragm
  - external intercostal muscles
- volumes

- capacities
- inspiratory reserve volume (IRV)
- tidal volume (V_T)
- expiratory reserve volume (ERV)
- inspiratory capacity
- vital capacity

- functional residual capacity (FRC)
- minimal volume
- total lung capacity
- residual volume

20.9

Pulmonary ventilation must be closely regulated to meet tissue oxygen demands

$$V_E = f \times V_T$$

- respiratory rate (f)
- respiratory minute volume (V_E)
- alveolar ventilation (V_A)
- anatomic dead space (V_D)

20.10

Gas diffusion depends on the partial pressures and solubilities of gases

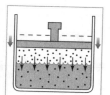

- gas laws
- partial pressure (P)
- Dalton's law

- Henry's law
  - external respiration
  - internal respiration

• = *Term boldfaced in this module*

20.11

Almost all of the oxygen in the blood is transported bound to hemoglobin within red blood cells

- heme units
- oxyhemoglobin (HbO_2)
- hemoglobin saturation
- oxygen-hemoglobin saturation curve
- Bohr effect
- 2,3-bisphosphoglycerate (BPG)

20.12

Most carbon dioxide transport occurs through the reversible formation of carbonic acid

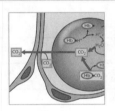

- carbaminohemoglobin ($HbCO_2$)
- carbonic anhydrase
- HbH^+
- chloride shift

20.13

Pulmonary disease can affect both lung elasticity and airflow

- compliance
- respiratory distress syndrome
- resistance
- chronic obstructive pulmonary disease (COPD)
- asthma
- chronic bronchitis
- blue bloaters
- emphysema
- pink puffers

• = *Term boldfaced in this module*

20.14

Respiratory control mechanisms involve interacting centers in the brain stem

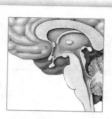

- respiratory rhythmicity centers
- dorsal respiratory group (DRG)
- ventral respiratory group (VRG)
- inspiratory center
- apneustic centers
- pneumotaxic centers
- higher centers
- quiet breathing
- forced breathing
- respiratory reflexes

20.15

Respiratory reflexes provide rapid automatic adjustments in pulmonary ventilation

HOMEOSTASIS

Normal arterial P_{CO_2}

- chemoreceptor reflexes
- hypercapnia
- hyperventilation
- hypocapnia
- shallow water blackout
- baroreceptor reflexes
- inflation reflex
- deflation reflex
- protective reflexes
- apnea

20.16

Respiratory function decreases with age; smoking makes matters worse

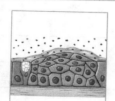

- lung cancers
- dysplasia
- metaplasia
- neoplasia
- neoplasm
- anaplasia

• = *Term boldfaced in this module*

1. Labeling

Label each of the structures of the digestive tract in the following figure.

a _____

b _____

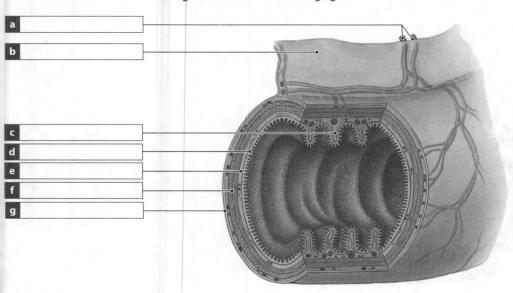

c _____

d _____

e _____

f _____

g _____

2. Matching

Match the following terms with the most closely related description.

- lamina propria
- peristalsis
- pacesetter cells
- esophagus
- muscularis mucosa
- segmentation
- plicae circulares
- sphincter
- myenteric plexus
- visceral smooth muscle cells
- plasticity
- liver
- multi-unit smooth muscle cells
- bolus

a _____ Digestive tube between the pharynx and stomach

b _____ Moves plicae circulares and villi

c _____ Areolar tissue layer containing blood vessels, nerve endings, and lymphatics

d _____ Permanent transverse folds in the digestive tract lining

e _____ Waves of muscular contractions that propel materials along digestive tract

f _____ Stimulate rhythmic cycles of activity along digestive tract

g _____ Nerve network within the muscularis externa

h _____ Have direct contact with motor neurons

i _____ Digestive system accessory organ

j _____ Rhythmic muscular contractions that mix materials in digestive tract

k _____ Form of food entering the digestive tract

l _____ Lack direct contact with motor neurons

m _____ Ability of smooth muscle cells to function over varied lengths

n _____ Ring of muscle tissue

3. Section integration

How would a decrease in smooth muscle tone affect the digestive processes and possibly promote constipation (infrequent bowel movement)?

_____

_____

_____

1. Labeling

Label each of the structures of the digestive tract in the following figure.

a _____

b _____

c _____

d _____

e _____

f _____

g _____

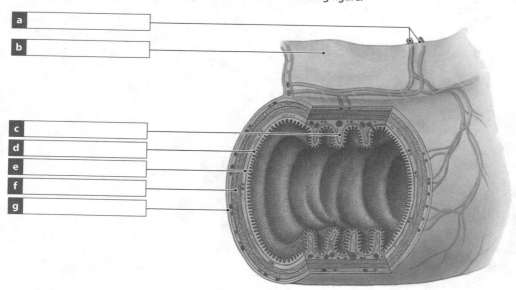

2. Matching

Match the following terms with the most closely related description.

- lamina propria
- peristalsis
- pacesetter cells
- esophagus
- muscularis mucosa
- segmentation
- plicae circulares
- sphincter
- myenteric plexus
- visceral smooth muscle cells
- plasticity
- liver
- multi-unit smooth muscle cells
- bolus

a _____ Digestive tube between the pharynx and stomach

b _____ Moves plicae circulares and villi

c _____ Areolar tissue layer containing blood vessels, nerve endings, and lymphatics

d _____ Permanent transverse folds in the digestive tract lining

e _____ Waves of muscular contractions that propel materials along digestive tract

f _____ Stimulate rhythmic cycles of activity along digestive tract

g _____ Nerve network within the muscularis externa

h _____ Have direct contact with motor neurons

i _____ Digestive system accessory organ

j _____ Rhythmic muscular contractions that mix materials in digestive tract

k _____ Form of food entering the digestive tract

l _____ Lack direct contact with motor neurons

m _____ Ability of smooth muscle cells to function over varied lengths

n _____ Ring of muscle tissue

3. Section integration

How would a decrease in smooth muscle tone affect the digestive processes and possibly promote constipation (infrequent bowel movement)?

_____

_____

_____

1. Labeling

Label the structures of a typical tooth in the following figure.

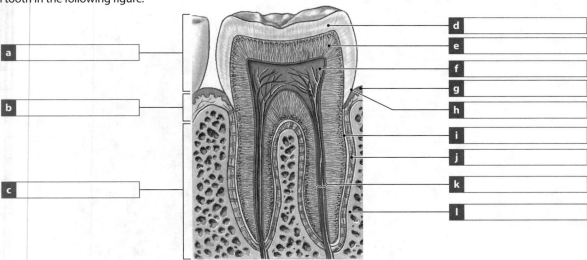

a	
b	
c	

d	
e	
f	
g	
h	
i	
j	
k	
l	

2 . Concept map

Using the following terms, fill in the blank boxes to complete the hormones of digestive activity concept map.

- acid production
- gastrin
- VIP
- insulin
- intestinal capillaries
- GIP
- material in jejunum
- gallbladder
- bile
- inhibits
- secretin and CCK
- nutrient utilization by tissues

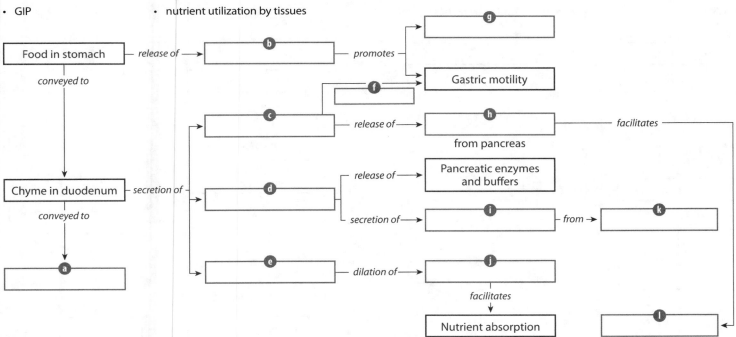

Hormones of Digestive Activity

Food in stomach — *release of* → [b] — *promotes* — [g]

[b] promotes → Gastric motility

Food in stomach — *conveyed to* →

[c] — *release of* → [h] from pancreas — *facilitates* →

[f] → Gastric motility

Chyme in duodenum — *secretion of* —

[d] — *release of* → Pancreatic enzymes and buffers

[d] — *secretion of* → [i] — *from* → [k]

Chyme in duodenum — *conveyed to* → [a]

[e] — *dilation of* → [j]

[j] — *facilitates* → Nutrient absorption

[l]

3. Short answer

Briefly describe the similarities and differences between parietal cells and chief cells in the stomach wall. _____

_____

_____

233

1. Labeling

Label the structures of a typical tooth in the following figure.

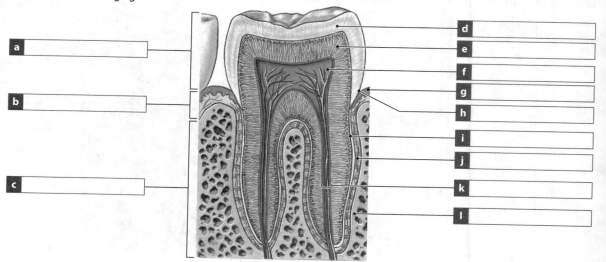

a

b

c

d

e

f

g

h

i

j

k

l

2 . Concept map

Using the following terms, fill in the blank boxes to complete the hormones of digestive activity concept map.

- acid production
- gastrin
- VIP
- insulin
- intestinal capillaries
- GIP
- material in jejunum
- gallbladder
- bile
- inhibits
- secretin and CCK
- nutrient utilization by tissues

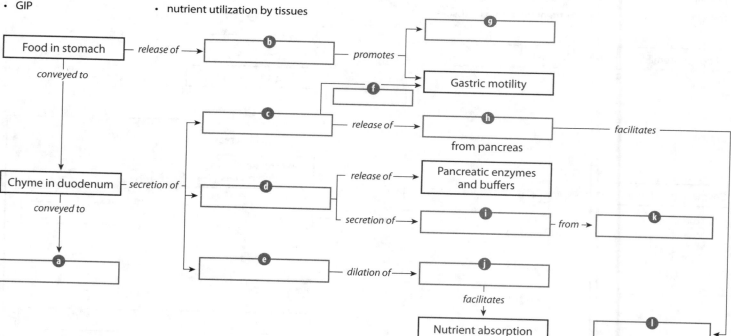

3. Short answer

Briefly describe the similarities and differences between parietal cells and chief cells in the stomach wall.

1. Labeling

Label each of the structures of a
liver lobule in the following diagram.

a

b

c

d

e

i

f

g

h

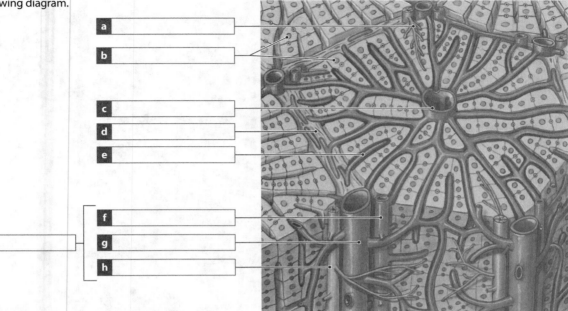

2. Matching

Match the following terms with the most closely related description.

- lysozyme
- emulsification
- gallstones
- Kupffer cells
- pancreatic lipase
- liver
- starch
- pancreas
- submandibular glands
- hepatocytes
- gallbladder
- mumps
- common bile duct
- peptic ulcer

a _____ Pancreatic alpha-amylase

b _____ Retroperitoneal organ

c _____ Drains liver and gallbladder

d _____ Bile-secreting cells

e _____ Viral infection of salivary glands

f _____ Digestive epithelial damage by acids

g _____ Process of breaking lipid droplets apart

h _____ Pancreatic enzyme that breaks down complex lipids

i _____ Organ that secretes bile continuously

j _____ Antibacterial enzyme

k _____ Greatest producer of saliva

l _____ Phagocytize and store iron

m _____ Stores bile

n _____ Cholecystitis

3. Short answer

Describe the beneficial roles of saliva. _____

_____

4. Section integration

Predict the consequences of a blockage of the duodenal ampulla by a tumor. _____

_____

1. Labeling

Label each of the structures of a
liver lobule in the following diagram.

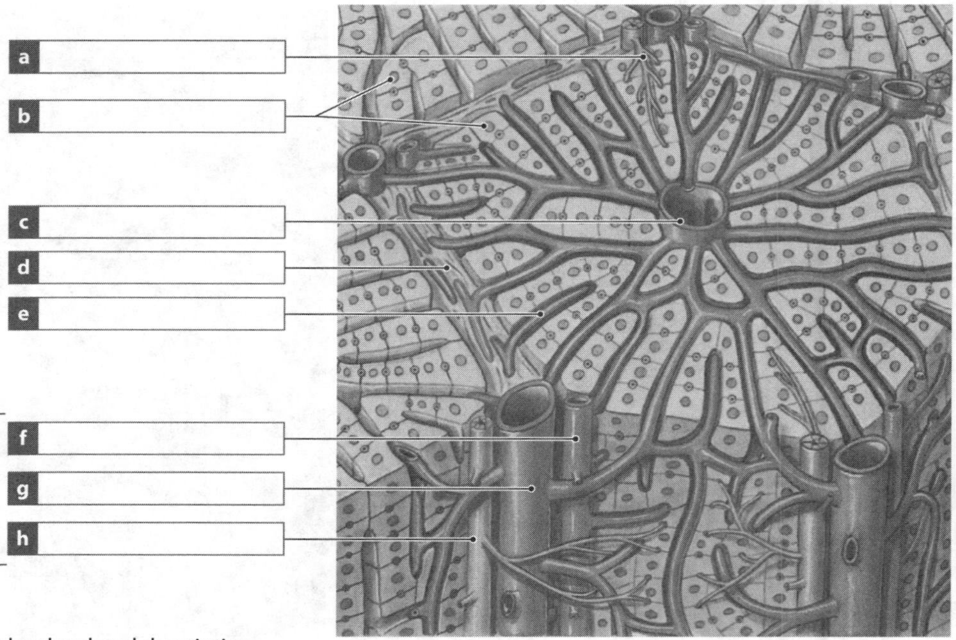

a _____

b _____

c _____

d _____

e _____

i _____

f _____

g _____

h _____

2. Matching

Match the following terms with the most closely related description.

- lysozyme
- emulsification
- gallstones
- Kupffer cells
- pancreatic lipase
- liver
- starch
- pancreas
- submandibular glands
- hepatocytes
- gallbladder
- mumps
- common bile duct
- peptic ulcer

a _____ Pancreatic alpha-amylase

b _____ Retroperitoneal organ

c _____ Drains liver and gallbladder

d _____ Bile-secreting cells

e _____ Viral infection of salivary glands

f _____ Digestive epithelial damage by acids

g _____ Process of breaking lipid droplets apart

h _____ Pancreatic enzyme that breaks down complex lipids

i _____ Organ that secretes bile continuously

j _____ Antibacterial enzyme

k _____ Greatest producer of saliva

l _____ Phagocytize and store iron

m _____ Stores bile

n _____ Cholecystitis

3. Short answer

Describe the beneficial roles of saliva. _____

_____

4. Section integration

Predict the consequences of a blockage of the duodenal ampulla by a tumor. _____

_____

Visual Outline with Key Terms

Summarize the content of each module using the terms in the order provided.

SECTION 1

General Organization of the Digestive System

- digestive system
- digestive tract
- gastrointestinal (GI) tract

_____

_____

_____

_____

_____

_____

_____

_____

21.1

The digestive tract is a muscular tube lined by a mucous epithelium

- mesentery
- mucosa
- submucosa
- muscularis externa
- serosa
- adventitia
- secretory glands

- plicae circulares
- villi
- muscularis mucosae
- submucosal plexus
- myenteric plexus

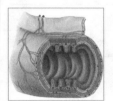

_____

_____

_____

_____

_____

_____

_____

_____

_____

21.2

Smooth muscle tissue is found throughout the body, but it plays a particularly prominent role in the digestive tract

- dense bodies
- multi-unit smooth muscle cells
- visceral smooth muscle cells

- pacesetter cells
- plasticity
- smooth muscle tone

_____

_____

_____

_____

_____

_____

_____

_____

 = _Term boldfaced in this module_

21.3

Smooth muscle contractions mix the contents of the digestive tract and propel materials along its length

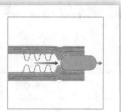

- bolus
- peristalsis
- segmentation
- myenteric reflexes
- enteroendocrine cells

SECTION 2

The Digestive Tract

- oral cavity, teeth, tongue
- pharynx
- esophagus
- stomach
- small intestine
- large intestine
- ingestion
- mechanical processing
- digestion
- secretion
- absorption
- compaction
- feces
- defecation

21.4

The oral cavity contains the tongue, salivary glands, and teeth, and receives secretions of the salivary glands

- oral cavity
- oral mucosa
- hard palate
- soft palate
- uvula
- root (of the tongue)
- labia
- cheeks
- body (of the tongue)
- vestibule
- labial frenulum
- lingual frenulum
- pharyngeal arches
- fauces
- tongue
- lingual lipase
- gingivae

• = *Term boldfaced in this module*

21.5

Teeth in different regions of the jaws vary in size, shape, and function

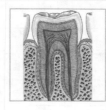

- dentin
- pulp cavity
- crown
- neck
- root
- alveolus
- occlusal surface
- enamel
- gingival sulcus
- cementum
- periodontal ligament
- gomphosis
- root canal
- apical foramen
- incisors
- cuspids
- bicuspids
- molars
- deciduous teeth
- primary dentition
- secondary dentition
- wisdom teeth
- dental arcades
- gingivitis
- tooth decay
- dental plaque

_____

_____

_____

_____

_____

_____

_____

_____

_____

_____

_____

21.6

The muscular walls of the pharynx and esophagus play a key role in swallowing

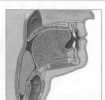

- pharynx
- esophagus
- adventitia
- deglutition
- buccal phase
- pharyngeal phase
- esophageal phase
- secondary peristaltic waves
- esophageal hiatus
- upper esophageal sphincter
- lower esophageal sphincter

_____

_____

_____

_____

_____

_____

_____

_____

21.7

The stomach and most of the intestinal tract are suspended by mesenteries within the peritoneal cavity

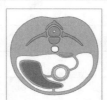

- peritoneal cavity
- visceral peritoneum
- parietal peritoneum
- dorsal and ventral mesenteries
- greater omentum
- lesser omentum
- falciform ligament
- mesentery proper
- mesocolon
- ascites

_____

_____

_____

_____

_____

_____

_____

_____

• = _Term boldfaced in this module_

21.8

The stomach is a muscular, expandable, J-shaped organ with three layers in the muscularis externa

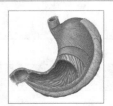

- lesser curvature
- greater curvature
- fundus
- cardia
- body
- pylorus
- pyloric antrum
- pyloric canal
- pyloric sphincter
- rugae

21.9

The stomach breaks down the organic nutrients in ingested materials

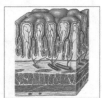

- mucosa
- submucosa
- muscularis externa
- serosa
- gastric glands
- parietal cells
- chief cells
- gastric pits
- intrinsic factor
- G cells
- pepsinogen
- pepsin
- rennin
- gastric lipase
- alkaline tide

21.10

The intestinal tract is specialized for the absorption of nutrients

- plicae circulares
- intestinal villi
- intestinal glands
- lacteal
- brush border

• = *Term boldfaced in this module*

21.11

The small intestine is divided into the duodenum, jejunum, and ileum

- duodenum
- jejunum
- ileum
- ileocecal valve
- plicae circulares
- duodenal glands

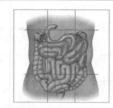

_____
_____
_____
_____
_____
_____

21.12

Five hormones are involved in the regulation of digestive activities

- gastrin
- secretin
- gastric inhibitory peptide (GIP)
- cholecystokinin (CCK)
- vasoactive intestinal peptide (VIP)

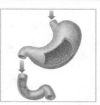

_____
_____
_____
_____
_____
_____

21.13

Central and local mechanisms coordinate gastric and intestinal activities

- phases of gastric secretion
- cephalic phase
- gastric phase
- mixing waves
- intestinal phase
- enterogastric reflex
- gastroenteric reflex
- gastroileal reflex
- ileocecal valve

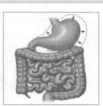

_____
_____
_____
_____
_____
_____

• = _Term boldfaced in this module_

21.14

The large intestine stores and concentrates fecal material

- large intestine
- cecum
- compaction
- ileocecal valve
- appendix
- appendicitis
- colon
- ascending colon
- right colic flexure
- transverse colon
- left colic flexure
- fatty appendices
- descending colon
- sigmoid flexure
- taeniae coli
- haustra
- sigmoid colon
- rectum
- mass movements

_____

_____

21.15

The large intestine compacts fecal material; the defecation reflex coordinates the elimination of feces

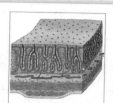

- hemorrhoids
- anal canal
- anal columns
- internal anal sphincter
- external anal sphincter
- anus
- indole
- skatole
- hydrogen sulfide
- defecation reflex
  ○ long reflex
  ○ short reflex

_____

_____

SECTION 3

Accessory Digestive Organs

○ salivary glands
○ gallbladder
○ pancreas
○ liver

_____

_____

● = *Term boldfaced in this module*

21.16

The salivary glands lubricate and moisten the mouth and initiate the digestion of complex carbohydrates

- sublingual salivary glands
- submandibular salivary glands
- salivary amylase
- parotid salivary glands
- parotid duct
- sublingual ducts
- submandibular duct
- saliva
- serous cells
- lysozyme
- mucous cells

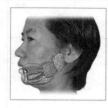

21.17

The liver, the largest visceral organ, is divided into left, right, caudate, and quadrate lobes

- liver
- falciform ligament
- left lobe
- right lobe
- porta hepatis
- coronary ligament
- bare area
- round ligament
- caudate lobe
- quadrate lobe
- gallbladder
- common bile duct

21.18

The liver tissues have an extensive and complex blood supply

- liver lobules
- hepatocytes
- portal areas
- liver sinusoids
- interlobular septum
- portal triad
- Kupffer cells
- central vein
- bile
- bile canaliculi
- bile ductules
- bile ducts
- viral hepatitis
- portal hypertension

● = *Term boldfaced in this module*

The gallbladder stores and concentrates bile, and the pancreas has vital endocrine and exocrine functions

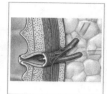

- gallbladder
- fundus, body, and neck of gallbladder
- right and left hepatic ducts
- common hepatic duct
- cystic duct
- common bile duct
- duodenal ampulla
- duodenal papilla
- hepatopancreatic sphincter
- emulsification
- pancreas
- head (of the pancreas)
- pancreatic duct
- pancreatic juice
- pancreatic lobules
- accessory pancreatic duct
- pancreatic acini
- pancreatic acinar cells
- pancreatic alpha-amylase
- pancreatic lipase
- nucleases
- proteolytic enzymes

Disorders of the digestive system are diverse and relatively common

- periodontal disease
- gingivitis
- mumps virus
- mumps
- esophagitis
- gastroesophageal reflux
- hepatitis
- cirrhosis
- hepatitis A, B, and C
- jaundice
- gallstones
- cholecystitis
- gastritis
- peptic ulcer
- gastric ulcer
- duodenal ulcer
- cimetidine
- pancreatitis
- enteritis
- diarrhea
- dysentery
- gastroenteritis
- colitis
- constipation
- colorectal cancer
- polyps

• = *Term boldfaced in this module*

1. Matching

Use the following terms to fill in the blanks in the cellular metabolism figure to the right.

- glucose
- electron transport system
- O_2
- fatty acids
- proteins
- citric acid cycle
- ATP
- CO_2
- H_2O
- coenzymes
- two-carbon chains

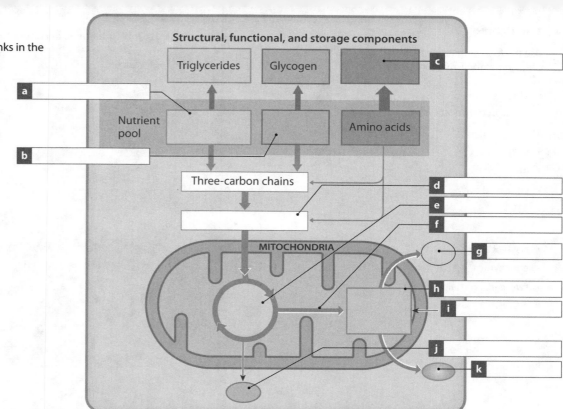

Structural, functional, and storage components

Triglycerides | Glycogen | [c]

[a]
Nutrient pool | | Amino acids

[b]

Three-carbon chains | [d]
[e]
[f]
MITOCHONDRIA
[g]
[h]
[i]
[j]
[k]

2. Short answer

Neural tissue requires a constant supply of glucose. What general shifts in cellular metabolism occur during fasting or starvation to meet that requirement?

_____

_____

_____

3. Matching

Match each of the following terms with the most closely related item.

- coenzymes
- cytochromes
- citric acid
- nutrients scarce
- nutrient pool
- anabolism
- ATP
- water
- acetate
- oxygen
- citric acid cycle
- catabolism
- oxidative phosphorylation
- nutrients abundant

a	_____	6-carbon molecule
b	_____	Collection of all the cell's organic substances
c	_____	Synthesis of new organic molecules
d	_____	Process that produces over 90 percent of ATP used by body cells
e	_____	ETS proteins
f	_____	Shuttle hydrogen atoms to the ETS
g	_____	Final acceptor of electrons from the ETS
h	_____	Breakdown of organic molecules
i	_____	Condition when cells preferentially break down carbohydrates
j	_____	Product of hydrogen ion diffusion within mitochondria
k	_____	Source of mitochondrial CO_2 production
l	_____	By-product of the ETS
m	_____	Condition when cells preferentially break down lipids
n	_____	Common substrate for mitochondrial ATP production

1. Matching

Use the following terms to fill in the blanks in the cellular metabolism figure to the right.

- glucose
- electron transport system
- O_2
- fatty acids
- proteins
- citric acid cycle
- ATP
- CO_2
- H_2O
- coenzymes
- two-carbon chains

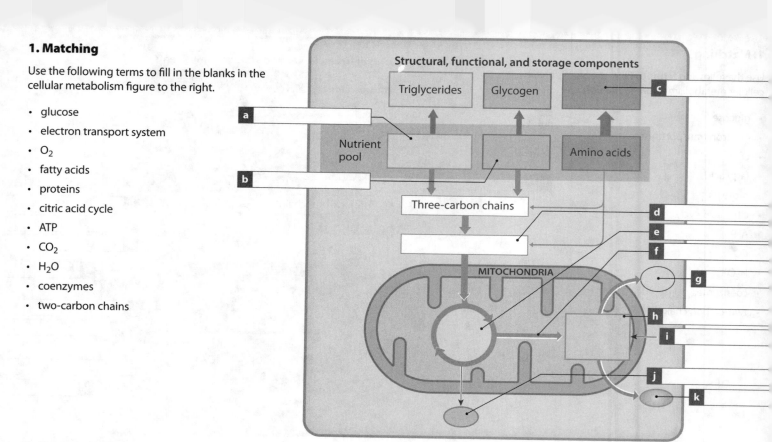

Structural, functional, and storage components

Triglycerides Glycogen c

Nutrient pool a b

Amino acids

Three-carbon chains d

e

f

MITOCHONDRIA

g

h

i

j

k

2. Short answer

Neural tissue requires a constant supply of glucose. What general shifts in cellular metabolism occur during fasting or starvation to meet that requirement?

_____

_____

_____

3. Matching

Match each of the following terms with the most closely related item.

- coenzymes
- cytochromes
- citric acid
- nutrients scarce
- nutrient pool
- anabolism
- ATP
- water
- acetate
- oxygen
- citric acid cycle
- catabolism
- oxidative phosphorylation
- nutrients abundant

a _____ 6-carbon molecule
b _____ Collection of all the cell's organic substances
c _____ Synthesis of new organic molecules
d _____ Process that produces over 90 percent of ATP used by body cells
e _____ ETS proteins
f _____ Shuttle hydrogen atoms to the ETS
g _____ Final acceptor of electrons from the ETS
h _____ Breakdown of organic molecules
i _____ Condition when cells preferentially break down carbohydrates
j _____ Product of hydrogen ion diffusion within mitochondria
k _____ Source of mitochondrial CO_2 production
l _____ By-product of the ETS
m _____ Condition when cells preferentially break down lipids
n _____ Common substrate for mitochondrial ATP production

1. Matching

Match each of the following terms with the most closely related item.

- lipogenesis
- anorexia
- lipolysis
- A, D, E, K
- absorptive state
- deamination
- ketone bodies
- calorie
- uric acid
- B complex and C
- urea formation
- lipoproteins
- insulin
- skeleton muscle

a	_____	Absorptive state hormone
b	_____	Glycogen reserves
c	_____	Water-soluble vitamins
d	_____	Fat catabolism
e	_____	Lipid synthesis
f	_____	Amino acid catabolism
g	_____	Fat-soluble vitamins
h	_____	Lipid transport
i	_____	Removal of amino group
j	_____	Lipid breakdown
k	_____	Gout
l	_____	Unit of energy
m	_____	Period following a meal
n	_____	Lack or loss of appetite

2. Multiple choice

Choose the bulleted item that best completes each statement.

a Intestinal absorption of nutrients occurs in the _____ .

- duodenum
- ileocecum
- ileum
- jejunum

b When blood glucose concentrations are elevated, the glucose molecules are _____ .

- catabolized for energy
- used to build proteins
- used for tissue repair
- all of these

c Most of the lipids absorbed by the digestive tract are immediately transferred to the _____ .

- liver
- red blood cells
- hepatocytes for storage
- venous circulation by the thoracic duct

d Hypervitaminosis involving water-soluble vitamins is relatively uncommon because _____ .

- the excess amount is stored in adipose tissue
- the excess amount is readily excreted in the urine
- the excess amount is stored in the bones
- excess amounts are readily absorbed by skeletal muscle tissue

3. Short answer

a What is the difference between an essential amino acid and a non-essential amino acid? _____
_____
_____
_____

b Describe four reasons why protein catabolism is an impractical source of quick energy. _____
_____
_____
_____

c What is the primary difference between the absorptive and postabsorptive states? _____
_____
_____
_____

d Why is the liver the focal point for metabolic regulation and control? _____
_____
_____
_____
_____

4. Section integration

Darla suffers from anorexia nervosa. One afternoon she is rushed to the emergency room because of cardiac arrhythmias. Her breath has the smell of an aromatic hydrocarbon, and blood and urine samples contain high levels of ketone bodies. Why do you think she is having the arrhythmias?

_____
_____

1. Matching

Match each of the following terms with the most closely related item.

• lipogenesis	a	_____	Absorptive state hormone
• anorexia	b	_____	Glycogen reserves
• lipolysis	c	_____	Water-soluble vitamins
• A, D, E, K	d	_____	Fat catabolism
• absorptive state	e	_____	Lipid synthesis
• deamination	f	_____	Amino acid catabolism
• ketone bodies	g	_____	Fat-soluble vitamins
• calorie	h	_____	Lipid transport
• uric acid	i	_____	Removal of amino group
• B complex and C	j	_____	Lipid breakdown
• urea formation	k	_____	Gout
• lipoproteins	l	_____	Unit of energy
• insulin	m	_____	Period following a meal
• skeleton muscle	n	_____	Lack or loss of appetite

2. Multiple choice

Choose the bulleted item that best completes each statement.

a Intestinal absorption of nutrients occurs in the _____.

- duodenum
- ileocecum
- ileum
- jejunum

b When blood glucose concentrations are elevated, the glucose molecules are _____.

- catabolized for energy
- used to build proteins
- used for tissue repair
- all of these

c Most of the lipids absorbed by the digestive tract are immediately transferred to the _____.

- liver
- red blood cells
- hepatocytes for storage
- venous circulation by the thoracic duct

d Hypervitaminosis involving water-soluble vitamins is relatively uncommon because _____.

- the excess amount is stored in adipose tissue
- the excess amount is readily excreted in the urine
- the excess amount is stored in the bones
- excess amounts are readily absorbed by skeletal muscle tissue

3. Short answer

a What is the difference between an essential amino acid and a non-essential amino acid? _____

_____

_____

_____

b Describe four reasons why protein catabolism is an impractical source of quick energy. _____

_____

_____

_____

c What is the primary difference between the absorptive and postabsorptive states? _____

_____

_____

_____

d Why is the liver the focal point for metabolic regulation and control? _____

_____

_____

_____

4. Section integration

Darla suffers from anorexia nervosa. One afternoon she is rushed to the emergency room because of cardiac arrhythmias. Her breath has the smell of an aromatic hydrocarbon, and blood and urine samples contain high levels of ketone bodies. Why do you think she is having the arrhythmias?

_____

_____

1. Matching

Match each of the following terms with the most closely related item.

- ghrelin
- basal metabolic rate
- 40 percent
- leptin
- insensible perspiration
- inhibits feeding center
- shivering thermogenesis
- peripheral vasoconstriction
- thermoregulation
- sensible perspiration
- neuropeptide Y
- 60 percent
- nonshivering thermogenesis
- peripheral vasodilation

a _____ Sweat gland activity; heat loss
b _____ Adipose tissue hormone
c _____ Homeostatic control of body temperature
d _____ General role of satiety center
e _____ Release of hormones; increased metabolism
f _____ Percent of catabolic energy released as heat
g _____ Stimulation of vasomotor center
h _____ Appetite-regulating neurotransmitter
i _____ Resting energy expenditure
j _____ Stomach hormone
k _____ Percent of catabolic energy captured as ATP
l _____ Inhibition of vasomotor center
m _____ Epithelial water loss
n _____ Result of increased skeletal muscle tone

2. Multiple choice

Choose the bulleted item that best completes each statement.

a An individual's BMR is influenced by their
_____ .

- gender
- body weight
- age
- all of these

b The four processes involved in heat exchange with the environment are _____ .

- sensible, insensible, heat loss, and heat gain
- radiation, conduction, convection, and evaporation
- physiological responses and behavioral modifications
- sensible, insensible, hormones, and heat conservation

c The primary mechanisms for increasing heat loss from the body include _____ .

- vasomotor and respiratory
- sensible and insensible
- physiological responses and behavioral modifications
- acclimatization and vasomotor

d All of the following are responses to an increase in body temperature, except _____ .

- stimulation of the respiratory centers
- stimulation of sweat glands
- peripheral vasoconstriction
- peripheral vasodilation

e If daily intake exceeds total energy demands, the excess energy is stored primarily as _____ .

- triglycerides in adipose tissue
- lipoproteins in the liver
- glycogen in the liver
- glucose in the bloodstream

f All of the following factors suppress appetite, except _____ .

- low blood glucose levels
- high blood glucose levels
- leptin
- stimulation of stretch receptors along the digestive tract

3. Short answer

a Why can energy consumption at rest be estimated by monitoring oxygen utilization? _____
_____
_____

b Describe the responses generated by the heat-gain center. _____
_____
_____

c Describe the heat-gain mechanisms involved in nonshivering thermogenesis. _____
_____
_____
_____
_____
_____

1. Matching

Match each of the following terms with the most closely related item.

- ghrelin   **a** _____ Sweat gland activity; heat loss
- basal metabolic rate **b** _____ Adipose tissue hormone
- 40 percent   **c** _____ Homeostatic control of body temperature
- leptin   **d** _____ General role of satiety center
- insensible perspiration **e** _____ Release of hormones; increased metabolism
- inhibits feeding center **f** _____ Percent of catabolic energy released as heat
- shivering thermogenesis **g** _____ Stimulation of vasomotor center
- peripheral vasoconstriction **h** _____ Appetite-regulating neurotransmitter
- thermoregulation   **i** _____ Resting energy expenditure
- sensible perspiration **j** _____ Stomach hormone
- neuropeptide Y   **k** _____ Percent of catabolic energy captured as ATP
- 60 percent   **l** _____ Inhibition of vasomotor center
- nonshivering thermogenesis **m** _____ Epithelial water loss
- peripheral vasodilation **n** _____ Result of increased skeletal muscle tone

2. Multiple choice

Choose the bulleted item that best completes each statement.

a An individual's BMR is influenced by their

_____ .

- gender
- body weight
- age
- all of these

b The four processes involved in heat exchange with the environment are _____ .

- sensible, insensible, heat loss, and heat gain
- radiation, conduction, convection, and evaporation
- physiological responses and behavioral modifications
- sensible, insensible, hormones, and heat conservation

c The primary mechanisms for increasing heat loss from the body include _____ .

- vasomotor and respiratory
- sensible and insensible
- physiological responses and behavioral modifications
- acclimatization and vasomotor

d All of the following are responses to an increase in body temperature except _____ .

- stimulation of the respiratory centers
- stimulation of sweat glands
- peripheral vasoconstriction
- peripheral vasodilation

e If daily intake exceeds total energy demands, the excess energy is stored primarily as _____ .

- triglycerides in adipose tissue
- lipoproteins in the liver
- glycogen in the liver
- glucose in the bloodstream

f All of the following factors suppress appetite, except

_____ .

- low blood glucose levels
- high blood glucose levels
- leptin
- stimulation of stretch receptors along the digestive tract

3. Short answer

a Why can energy consumption at rest be estimated by monitoring oxygen utilization? _____

_____

_____

b Describe the responses generated by the heat-gain center. _____

_____

_____

_____

c Describe the heat-gain mechanisms involved in nonshivering thermogenesis. _____

_____

_____

_____

_____

_____

_____

Visual Outline with Key Terms

Summarize the content of each module using the terms in the order provided.

SECTION 1

An Introduction to Cellular Metabolism

- metabolism
- cellular metabolism
- catabolism
- anabolism
- metabolic turnover
- nutrient pool

NUTRIENT POOL

ATP

Aerobic Metabolism
(in mitochondria)

22.1

Cells obtain most of their ATP from the electron transport system, which is linked to the citric acid cycle

- acetate
- acetyl-CoA
- acetyl group
- citric acid
- citric acid cycle
- NAD
- FAD
- oxidative phosphorylation
- electron transport system (ETS)
- cytochromes
- ATP synthase

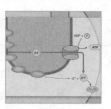

ADP + P

ATP

22.2

Cells can break down any available substrate from the nutrient pool to obtain the energy they need

- nutrient pool
- catabolic pathway
- anabolic pathway
- glycogenesis
- glycogenolysis
- gluconeogenesis
- glycolysis

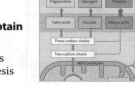

Structural, functional, and storage components

| Triglycerides | Glycogen | Proteins |

| Fatty acids | Glucose | Amino acids |

Three-carbon chains

Two-carbon chains

MITOCHONDRIA

SECTION 2

The Digestion and Metabolism of Organic Nutrients

- oral cavity
- stomach
- duodenum
- jejunum
- hepatic portal vein
- liver

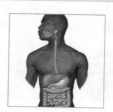

22.3

Carbohydrates are usually the preferred substrates for catabolism and ATP production under resting conditions

- salivary amylase
- pancreatic alpha-amylase
- maltase
- sucrase
- lactase
- flatus
- liver
- skeletal muscle
- glucose
- glycogen
- insulin
- pyruvate

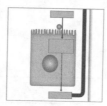

22.4

Glycolysis is the first step in the catabolism of the carbohydrate glucose

- anaerobic
- cytosol
- pyruvate
- glycogenolysis

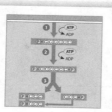

● = *Term boldfaced in this module*

22.5

Lipids reach the bloodstream in chylomicrons; the cholesterol is then extracted and released as lipoproteins

- lingual lipase
- pancreatic lipase
- bile salts
- emulsification
- micelles
- chylomicrons
- lipoproteins
- lacteals
- thoracic duct
- lipoprotein lipase
- cholesterol
- low-density lipoproteins (LDLs)
- high-density lipoproteins (HDLs)

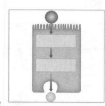

22.6

Fatty acids can be broken down to provide energy or converted to other lipids

- lipolysis
- beta-oxidation
- lipogenesis
- linolenic acid
- omega-3
- linoleic acid
- omega-6
- essential fatty acids

22.7

An amino acid not needed for protein synthesis may be broken down or converted to a different amino acid

- pepsin
- enteropeptidase
- trypsin
- chymotrypsin
- carboxypeptidase
- elastase
- peptidases
- dipeptidases
- hepatic portal vein
- liver
- essential amino acids
- amination
- ammonium ion
- transamination
- deamination
- urea
- urea cycle

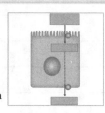

● = *Term boldfaced in this module*

22.8

There are two general patterns of metabolic activity: the absorptive and postabsorptive states

- absorptive state
  - insulin
  - growth hormone
  - androgens
  - estrogens
- postabsorptive state
- ketone bodies
  - glucocorticoids
  - growth hormone
  - glucagon
  - epinephrine

| Glucose levels elevated |
| Lipid levels elevated |
| Amino acids elevated |

22.9

Vitamins are essential to the function of many metabolic pathways

- nutrition
- vitamins
- fat-soluble vitamins
- avitaminosis
- hypervitaminosis
- water-soluble vitamins
- intrinsic factor

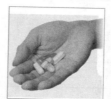

22.10

Proper nutrition depends on eating a balanced diet

- balanced diet
- malnutrition
  - food pyramid
- MyPyramid.gov Steps to a Healthier You
- calories
- joules
- kilocalorie
- kilojoule (kJ)
- Calorie
- complete proteins
- incomplete proteins

• = Term boldfaced in this module

22.11

Metabolic disorders may result from nutritional or biochemical problems

- eating disorders
- anorexia
- anorexia nervosa
- bulimia
- obesity
- regulatory obesity
- metabolic obesity
- ○ cholesterol
- ○ atherosclerosis

- ○ coronary artery disease
- phenylketonuria (PKU)
- protein deficiency disease
- kwashiorkor
- ketone bodies
- ketosis

- ketoacidosis
- uric acid
- nitrogenous wastes
- gout
- gouty arthritis

SECTION 3

Energetics and Thermoregulation

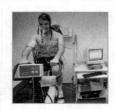

- energetics
- basal metabolic rate (BMR)
- thermoregulation

22.12

The control of appetite is complex and involves both short-term and long-term mechanisms

- feeding center
- satiety center
- neuropeptide Y (NPY)

- ghrelin
- leptin

• = *Term boldfaced in this module*

22.13

To maintain a constant body temperature, heat gain and heat loss must be in balance

- ○ heat transfer
- • radiation
- • convection
- • evaporation
- • insensible perspiration
- • sensible perspiration
- • conduction

22.14

Hypothalamic thermoregulatory centers adjust rates of heat gain and heat loss

- • heat-loss center
- • heat-gain center
- ○ preoptic area
- • behavioral changes
- • vasodilation and shunting of blood to skin surface
- ○ vasomotor center
- ○ radiation
- ○ convection
- • sweat production
- • respiratory heat loss
- • nonshivering thermogenesis
- • shivering thermogenesis
- • countercurrent exchange

• = *Term boldfaced in this module*

1. Short answer

Label the kidney structures in the following diagram, and then provide a brief functional/anatomical description of each.

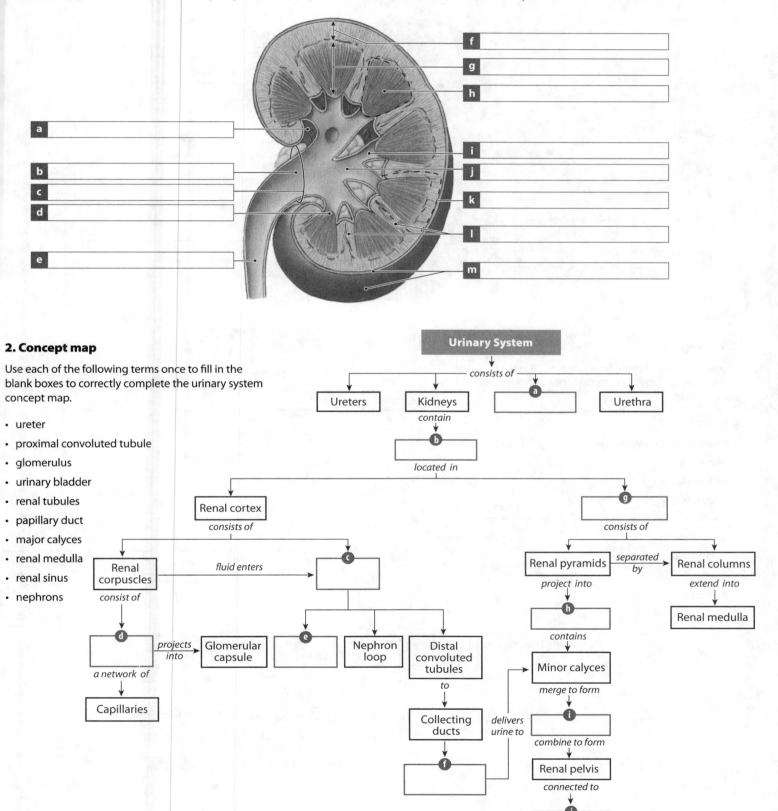

a	
b	
c	
d	
e	
f	
g	
h	
i	
j	
k	
l	
m	

2. Concept map

Use each of the following terms once to fill in the blank boxes to correctly complete the urinary system concept map.

- ureter
- proximal convoluted tubule
- glomerulus
- urinary bladder
- renal tubules
- papillary duct
- major calyces
- renal medulla
- renal sinus
- nephrons

Urinary System

consists of

Ureters   Kidneys   [a]   Urethra

contain

[b]

located in

Renal cortex

consists of

Renal corpuscles

fluid enters → [c]

consist of

[d]

projects into → Glomerular capsule   [e]   Nephron loop   Distal convoluted tubules

a network of

Capillaries

to

Collecting ducts

[f]

[g]

consists of

Renal pyramids — separated by → Renal columns

project into

extend into

[h]

Renal medulla

contains

delivers urine to

Minor calyces

merge to form

[i]

combine to form

Renal pelvis

connected to

[j]

1. Short answer

Label the kidney structures in the following diagram, and then provide a brief functional/anatomical description of each.

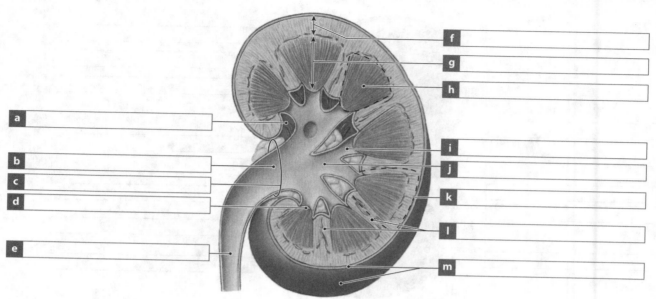

a

b

c

d

e

f

g

h

i

j

k

l

m

2. Concept map

Use each of the following terms once to fill in the blank boxes to correctly complete the urinary system concept map.

- ureter
- proximal convoluted tubule
- glomerulus
- urinary bladder
- renal tubules
- papillary duct
- major calyces
- renal medulla
- renal sinus
- nephrons

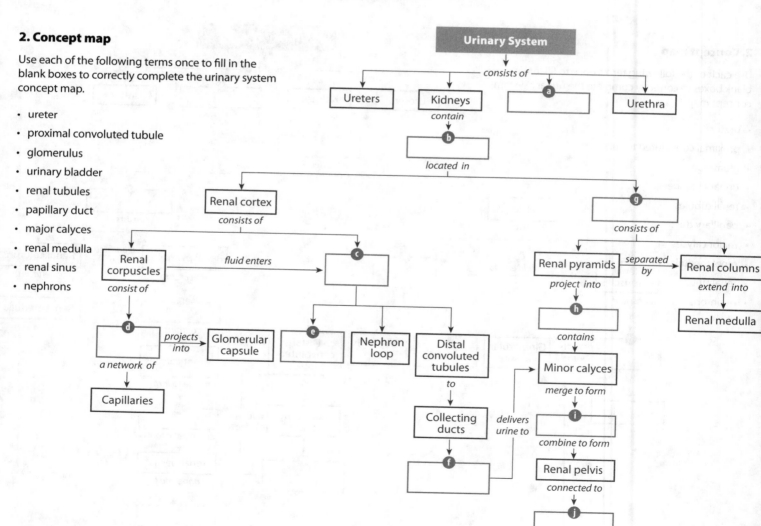

1. Short answer

Identify the structures of the representative nephron in the following diagram, and describe the functions of each.

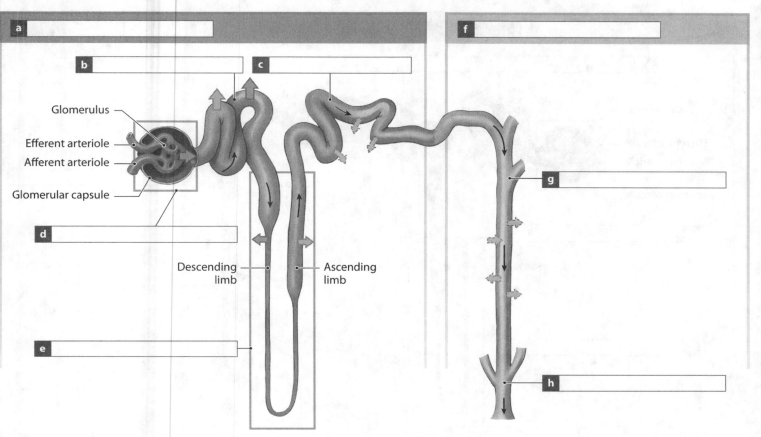

a [_____]

b [_____] c [_____]

f [_____]

Glomerulus

Efferent arteriole

Afferent arteriole

Glomerular capsule

d [_____]

Descending limb — Ascending limb

g [_____]

e [_____]

h [_____]

2. Matching

Match the following terms with their most closely related item.

- aquaporins
- ADH
- aldosterone
- PCT
- secretion
- renal corpuscle
- nephron loop
- filtrate
- BCOP
- podocytes

a [_____] Site of plasma filtration

b [_____] Glomerular epithelium

c [_____] Protein-free solution

d [_____] Opposes filtration

e [_____] Countercurrent multiplication

f [_____] Water channels

g [_____] Primary method for eliminating drugs or toxins

h [_____] Ion pump—Na^+ reabsorbed

i [_____] Primary site of nutrient reabsorption in the nephron

j [_____] Regulates passive reabsorption of water from urine in the collecting system

3. Section integration

Marissa has had a urinalysis that detected large amounts of plasma proteins and white blood cells in her urine.
What condition might be responsible, and what effects would it have on her urine output?

_____

_____

1. Short answer

Identify the structures of the representative nephron in the following diagram, and describe the functions of each.

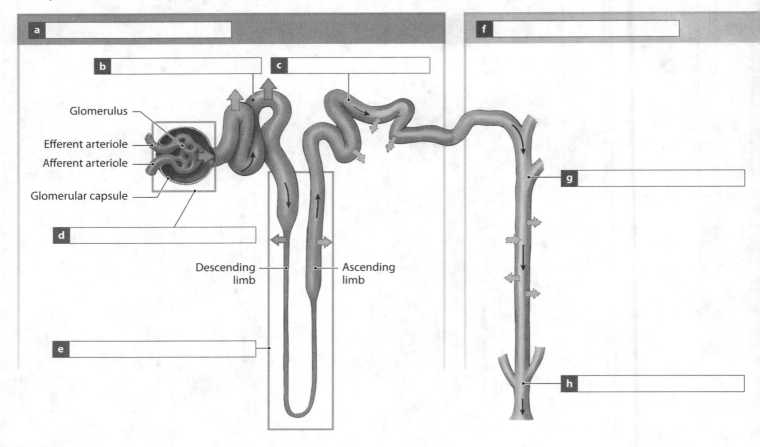

a _____

f _____

b _____

c _____

Glomerulus

Efferent arteriole

Afferent arteriole

Glomerular capsule

d _____

Descending limb

Ascending limb

g _____

e _____

h _____

2. Matching

Match the following terms with their most closely related item.

- aquaporins
- ADH
- aldosterone
- PCT
- secretion
- renal corpuscle
- nephron loop
- filtrate
- BCOP
- podocytes

a _____ Site of plasma filtration

b _____ Glomerular epithelium

c _____ Protein-free solution

d _____ Opposes filtration

e _____ Countercurrent multiplication

f _____ Water channels

g _____ Primary method for eliminating drugs or toxins

h _____ Ion pump—Na^+ reabsorbed

i _____ Primary site of nutrient reabsorption in the nephron

j _____ Regulates passive reabsorption of water from urine in the collecting system

3. Section integration

Marissa has had a urinalysis that detected large amounts of plasma proteins and white blood cells in her urine. What condition might be responsible, and what effects would it have on her urine output?

_____

_____

1. Matching

Match the following terms with their descriptions.

- urethra
- external urethral sphincter
- detrusor
- rugae
- internal urethral sphincter
- trigone
- transitional epithelium
- micturition
- stratified squamous epithelium
- external urethral orifice
- middle umbilical ligament

a _____ The ring of smooth muscle in the neck of the urinary bladder

b _____ Triangular area within the urinary bladder

c _____ Relaxation of this muscle leads to urination

d _____ The external opening of the urethra

e _____ Folds lining the surface of the empty urinary bladder

f _____ Superior, supporting fibrous cord of the urinary bladder

g _____ Contraction of this smooth muscle compresses the urinary bladder

h _____ The type of epithelium that lines the ureters

i _____ Tube that transports urine to the exterior

j _____ Term for urination

k _____ Epithelium that lines the urethra

2. Labeling

Use the following descriptions to fill in the boxes in the micturition reflex diagram below.

- sensation relayed to thalamus
- individual relaxes external urethral sphincter
- afferent fibers carry information to sacral spinal cord
- sensation of bladder fullness delivered to cerebral cortex
- stretch receptors stimulated
- detrusor muscle contraction stimulated
- parasympathetic preganglionic fibers carry motor commands
- internal urethral sphincter relaxes

3. Short answer

List four primary signs and symptoms of urinary disorders.

4. Short answer

Briefly describe the similarities and differences in the following pairs of terms.

- cystitis/pyelonephritis
- stress incontinence/ overflow incontinence
- polyuria/proteinuria

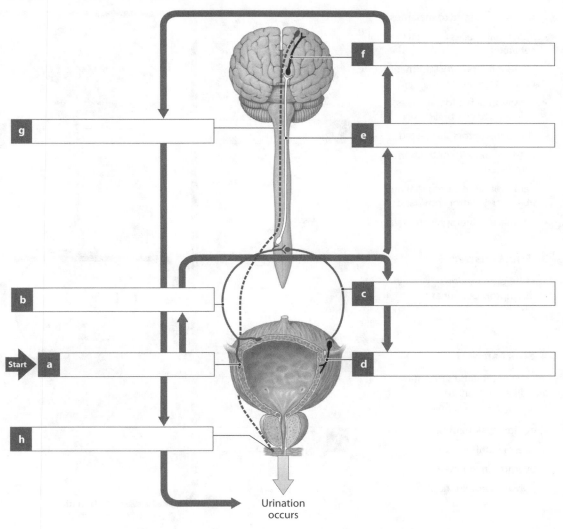

Urination occurs

1. Matching

Match the following terms with their descriptions.

- urethra
- external urethral sphincter
- detrusor
- rugae
- internal urethral sphincter
- trigone
- transitional epithelium
- micturition
- stratified squamous epithelium
- external urethral orifice
- middle umbilical ligament

a _____ The ring of smooth muscle in the neck of the urinary bladder

b _____ Triangular area within the urinary bladder

c _____ Relaxation of this muscle leads to urination

d _____ The external opening of the urethra

e _____ Folds lining the surface of the empty urinary bladder

f _____ Superior, supporting fibrous cord of the urinary bladder

g _____ Contraction of this smooth muscle compresses the urinary bladder

h _____ The type of epithelium that lines the ureters

i _____ Tube that transports urine to the exterior

j _____ Term for urination

k _____ Epithelium that lines the urethra

2. Labeling

Use the following descriptions to fill in the boxes in the micturition reflex diagram below.

- sensation relayed to thalamus
- individual relaxes external urethral sphincter
- afferent fibers carry information to sacral spinal cord
- sensation of bladder fullness delivered to cerebral cortex
- stretch receptors stimulated
- detrusor muscle contraction stimulated
- parasympathetic preganglionic fibers carry motor commands
- internal urethral sphincter relaxes

3. Short answer

List four primary signs and symptoms of urinary disorders.

4. Short answer

Briefly describe the similarities and differences in the following pairs of terms.

- cystitis/pyelonephritis
- stress incontinence/ overflow incontinence
- polyuria/proteinuria

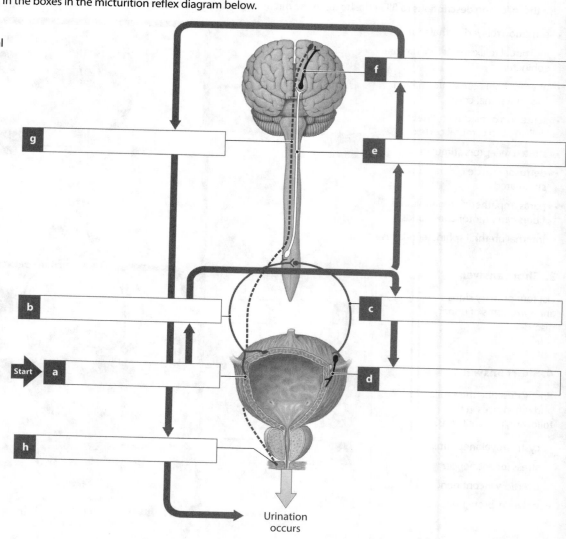

Urination occurs

Visual Outline with Key Terms

Summarize the content of each module using the terms in the order provided.

SECTION 1

Anatomy of the Urinary System

- urinary system
- urinary tract
- kidneys
- urine
- ureters
- urinary bladder
- urethra
- urination

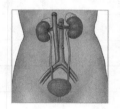

_____

_____

_____

_____

_____

_____

_____

23.1

The kidneys are paired retroperitoneal organs

- kidney
- hilum
- ureters
- retroperitoneal
- fibrous capsule
- perinephric fat capsule
- renal fascia

_____

_____

_____

_____

_____

_____

_____

_____

_____

23.2

The kidneys are complex at the gross and microscopic levels

- hilum
- fibrous capsule
- renal sinus
- renal cortex
- renal medulla
- renal pyramid
- renal papilla
- renal column
- kidney lobe
- minor and major calyces
- renal pelvis
- nephron
- cortical nephron
- juxtamedullary nephron
- nephron loop

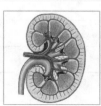

_____

_____

_____

_____

_____

_____

_____

_____

_____

_____

● = _Term boldfaced in this module_

23.3

A nephron can be divided into regions; each region has specific functions

- nephron
- renal corpuscle
- renal tubule
- capsular space
- filtrate
- glomerular capsule
- glomerulus
- proximal convoluted tubule (PCT)
- nephron loop
- distal convoluted tubule (DCT)
- collecting system
- collecting duct
- papillary duct

23.4

The kidneys are highly vascular, and the circulation patterns are complex

- renal artery and renal vein
- segmental arteries
- interlobar arteries and veins
- arcuate arteries and veins
- cortical radiate arteries and veins
- afferent and efferent arterioles
- glomerulus
- peritubular capillaries
- vasa recta

SECTION 2

Overview of Renal Physiology

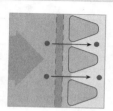

- urea
- creatinine
- uric acid
- filtration
- reabsorption
- secretion

• = *Term boldfaced in this module*

23.5

Filtration, reabsorption, and secretion occur in specific regions of the nephron and collecting system

- ○ filtration
- ○ water reabsorption
- ○ solute reabsorption
- ○ solute secretion

_____

_____

_____

_____

_____

_____

_____

23.6

Filtration occurs at the renal corpuscle

- • efferent arteriole
- • juxtaglomerular complex
- • afferent arteriole
- • capsular space
- • podocytes
- • pedicels
- • filtration slits
- • mesangial cells
- • dense layer
- • filtration membrane
- • hydrostatic pressure
- • colloid osmotic pressure
- • glomerular hydrostatic pressure (GHP)
- • blood colloid osmotic pressure (BCOP)
- • net filtration pressure (NFP)
- • capsular hydrostatic pressure (CsHP)
- • capsular colloid osmotic pressure

_____

_____

_____

_____

_____

_____

_____

_____

_____

_____

_____

_____

23.7

The glomerular filtration rate (GFR) is the amount of filtrate produced each minute

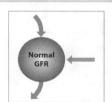

- • glomerular filtration rate (GFR)
- ○ filtrate
- • autoregulation
- • central regulation
- • endocrine response
- • neural responses
- ○ renin
- ○ angiotensin
- • angiotensin converting enzyme (ACE)
- ○ aldosterone
- ○ ADH

_____

_____

_____

_____

_____

_____

_____

_____

• = *Term boldfaced in this module*

23.8

Reabsorption predominates along the PCT, whereas reabsorption and secretion are often linked along the DCT

- proximal convoluted tubule (PCT)
- distal convoluted tubule (DCT)
- reabsorption
- secretion
- diffusion
- facilitated diffusion
- active transport
- secondary active transport
- peritubular capillary
- peritubular fluid
- tubular fluid

23.9

Feedback between the nephron loop and the collecting duct creates the osmotic gradient in the renal medulla

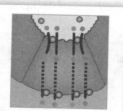

- nephron loop
- countercurrent multiplication
- active transport
- impermeable
- osmosis
- thick ascending loop
- thin descending loop
- renal medulla

23.10

Urine volume and concentration are hormonally regulated

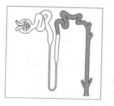

- obligatory water reabsorption
- facultative water reabsorption
- distal convoluted tubule
- collecting duct
- antidiuretic hormone (ADH)
- aquaporins
- urine

23.11

Renal function is an integrative process

- osmotic concentration
- obligatory water reabsorption
- urea
- aldosterone
- ADH
- vasa recta

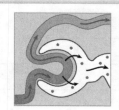

23.12

Renal failure is a life-threatening condition

- renal failure
- chronic renal failure
- acute renal failure
- hemodialysis
- dialysis
- dialysis fluid
- shunts

SECTION 3

Urine Storage and Elimination

- pyelogram
- ureters
- urinary bladder
- urethra

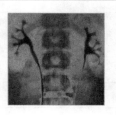

= Term boldfaced in this module

23.13

The ureters, urinary bladder, and urethra are specialized for the conduction of urine

- lateral umbilical ligaments
- middle umbilical ligament
- ureters
- ureteral openings
- trigone
- rugae
- neck of the urinary bladder
- internal urethral sphincter
- external urethral sphincter
- detrusor muscle

23.14

Urination involves a reflex coordinated by the nervous system

- micturition reflex
- local pathway
- central pathway
- internal urethral sphincter
- external urethral sphincter

23.15

Urinary disorders can often be detected by physical exams and laboratory tests

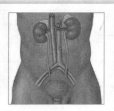

- pyelonephritis
- renal calculi
- dysuria
- polyuria
- oliguria
- anuria
- incontinence
- urinary retention
- edema
- proteinuria
- fever
- cystitis

● = *Term boldfaced in this module*

1. Matching

Match the following terms with the most closely related description.

- kidneys
- potassium
- fluid compartments
- fluid balance
- hypertonic plasma
- dehydration
- aldosterone
- plasma, interstitial fluid
- osmoreceptors
- fluid shift
- hypokalemia
- ADH
- hyponatremia
- sodium

a _____ Monitor plasma osmotic concentration

b _____ Water gain = water loss

c _____ Major components of ECF

d _____ Dominant cation in ECF

e _____ Hormone that restricts water loss and stimulates thirst

f _____ Overhydration

g _____ Dominant cation in ICF

h _____ ICF and ECF

i _____ Most important sites of sodium ion regulation

j _____ Water movement between ECF and ICF

k _____ Water moves from cells into ECF

l _____ Result of aldosteronism

m _____ Water losses greater than water gains

n _____ Regulates sodium ion absorption along DCT and collecting system

2. Multiple choice

Choose the bulleted item that best completes each statement.

a Nearly two-thirds of the total body water content is

_____ .

- extracellular fluid (ECF)
- intracellular fluid (ICF)
- tissue fluid
- interstitial fluid

b Electrolyte balance involves balancing the rates of absorption across the digestive tract with rates of loss at the _____ .

- heart and lungs
- stomach and liver
- kidneys and sweat glands
- pancreas and gallbladder

c If the ECF is hypertonic with respect to the ICF, water will move

_____ .

- from the ECF into the cells until osmotic equilibrium is restored
- from the cells into the ECF until osmotic equilibrium is restored
- in both directions until osmotic equilibrium is restored
- in response to the sodium–potassium exchange pump

d When pure water is consumed, the ECF

_____ .

- becomes hypotonic with respect to the ICF
- becomes hypertonic with respect to the ICF
- becomes isotonic with respect to the ICF
- electrolytes become more concentrated

e Physiological adjustments affecting fluid and electrolyte balance are mediated primarily by _____ .

- antidiuretic hormone
- aldosterone
- natriuretic peptides
- all of these

f When water is lost but electrolytes are retained, the osmolarity of the ECF rises, and osmosis then moves water _____ .

- out of the ECF and into the ICF
- back and forth between the ICF and ECF
- out of the ICF and into the ECF
- none of the above

3. Section integration

Malia, a nursing student, has been caring for burn patients. She notices that they consistently show elevated levels of potassium in their urine and wonders why. What would you tell her? _____

_____

_____

1 . Matching

Match the following terms with the most closely related description.

- kidneys
- potassium
- fluid compartments
- fluid balance
- hypertonic plasma
- dehydration
- aldosterone
- plasma, interstitial fluid
- osmoreceptors
- fluid shift
- hypokalemia
- ADH
- hyponatremia
- sodium

a	_____	Monitor plasma osmotic concentration
b	_____	Water gain = water loss
c	_____	Major components of ECF
d	_____	Dominant cation in ECF
e	_____	Hormone that restricts water loss and stimulates thirst
f	_____	Overhydration
g	_____	Dominant cation in ICF
h	_____	ICF and ECF
i	_____	Most important sites of sodium ion regulation
j	_____	Water movement between ECF and ICF
k	_____	Water moves from cells into ECF
l	_____	Result of aldosteronism
m	_____	Water losses greater than water gains
n	_____	Regulates sodium ion absorption along DCT and collecting system

2. Multiple choice

Choose the bulleted item that best completes each statement.

a Nearly two-thirds of the total body water content is

_____ .

- extracellular fluid (ECF)
- intracellular fluid (ICF)
- tissue fluid
- interstitial fluid

b Electrolyte balance involves balancing the rates of absorption across the digestive tract with rates of loss at the _____ .

- heart and lungs
- stomach and liver
- kidneys and sweat glands
- pancreas and gallbladder

c If the ECF is hypertonic with respect to the ICF, water will move

_____ .

- from the ECF into the cells until osmotic equilibrium is restored
- from the cells into the ECF until osmotic equilibrium is restored
- in both directions until osmotic equilibrium is restored
- in response to the sodium–potassium exchange pump

d When pure water is consumed, the ECF

_____ .

- becomes hypotonic with respect to the ICF
- becomes hypertonic with respect to the ICF
- becomes isotonic with respect to the ICF
- electrolytes become more concentrated

e Physiological adjustments affecting fluid and electrolyte balance are mediated primarily by _____ .

- antidiuretic hormone
- aldosterone
- natriuretic peptides
- all of these

f When water is lost but electrolytes are retained, the osmolarity of the ECF rises, and osmosis then moves water _____ .

- out of the ECF and into the ICF
- back and forth between the ICF and ECF
- out of the ICF and into the ECF
- none of the above

3. Section integration

Malia, a nursing student, has been caring for burn patients. She notices that they consistently show elevated levels of potassium in their urine and wonders why. What would you tell her? _____

_____

_____

1. Labeling

Use the following terms to label the boxes in the two flowcharts. Terms may be used more than once.

- plasma pH decrease
- plasma pH increase
- increased P_{CO_2}
- decreased P_{CO_2}
- increased
- decreased
- alkalosis
- acidosis
- generated
- secreted

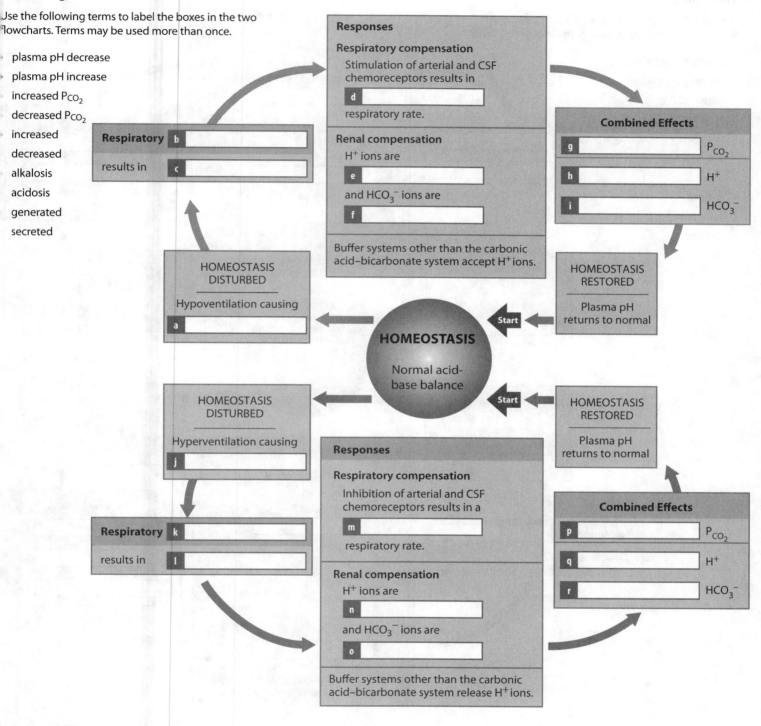

Responses

Respiratory compensation
Stimulation of arterial and CSF chemoreceptors results in **d** _____ respiratory rate.

Renal compensation
H$^+$ ions are **e** _____
and HCO$_3^-$ ions are **f** _____

Buffer systems other than the carbonic acid–bicarbonate system accept H$^+$ ions.

Combined Effects
g _____ P$_{CO_2}$
h _____ H$^+$
i _____ HCO$_3^-$

Respiratory **b** _____
results in **c** _____

HOMEOSTASIS DISTURBED
Hypoventilation causing
a _____

HOMEOSTASIS RESTORED
Plasma pH returns to normal

HOMEOSTASIS
Normal acid-base balance

Start

HOMEOSTASIS DISTURBED
Hyperventilation causing
j _____

HOMEOSTASIS RESTORED
Plasma pH returns to normal

Responses

Respiratory compensation
Inhibition of arterial and CSF chemoreceptors results in a **m** _____ respiratory rate.

Renal compensation
H$^+$ ions are **n** _____
and HCO$_3^-$ ions are **o** _____

Buffer systems other than the carbonic acid–bicarbonate system release H$^+$ ions.

Combined Effects
p _____ P$_{CO_2}$
q _____ H$^+$
r _____ HCO$_3^-$

Respiratory **k** _____
results in **l** _____

2. Section integration

After falling into an abandoned stone quarry filled with water and nearly drowning, a young boy is rescued. His rescuers assess his condition and find that his body fluids have high P_{CO_2} and lactate levels, and low P_{O_2} levels. Identify the underlying problem and recommend the necessary treatment to restore homeostatic conditions. _____

_____

_____

1. Labeling

Use the following terms to label the boxes in the two flowcharts. Terms may be used more than once.

- plasma pH decrease
- plasma pH increase
- increased P_{CO_2}
- decreased P_{CO_2}
- increased
- decreased
- alkalosis
- acidosis
- generated
- secreted

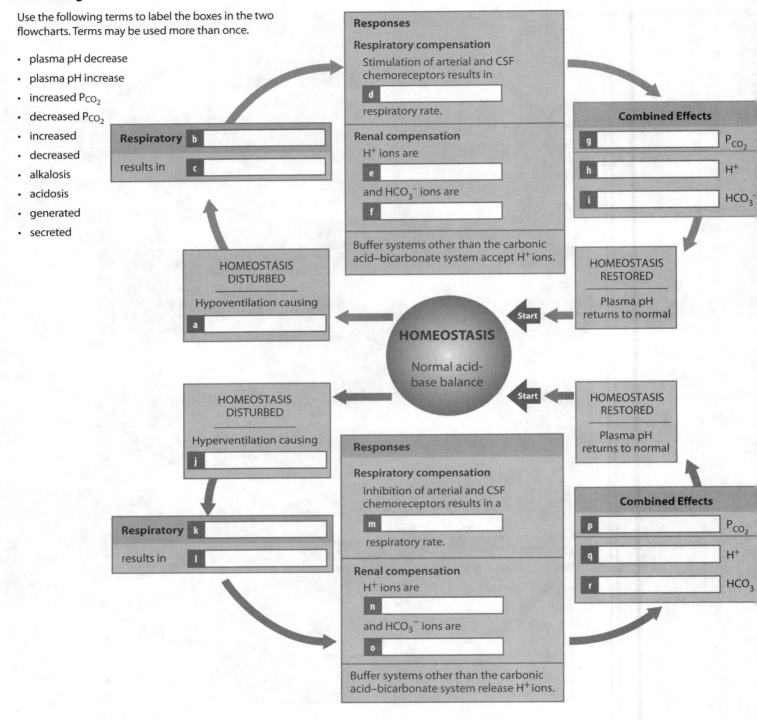

Responses

Respiratory compensation
Stimulation of arterial and CSF chemoreceptors results in
[d] _____ respiratory rate.

Renal compensation
H^+ ions are
[e] _____
and HCO_3^- ions are
[f] _____

Buffer systems other than the carbonic acid–bicarbonate system accept H^+ ions.

Combined Effects
[g] _____ P_{CO_2}
[h] _____ H^+
[i] _____ HCO_3^-

Respiratory [b] _____
results in [c] _____

HOMEOSTASIS DISTURBED
Hypoventilation causing
[a] _____

HOMEOSTASIS RESTORED
Plasma pH returns to normal

HOMEOSTASIS
Normal acid-base balance

Start

HOMEOSTASIS DISTURBED
Hyperventilation causing
[j] _____

Respiratory [k] _____
results in [l] _____

HOMEOSTASIS RESTORED
Plasma pH returns to normal

Responses

Respiratory compensation
Inhibition of arterial and CSF chemoreceptors results in a
[m] _____ respiratory rate.

Renal compensation
H^+ ions are
[n] _____
and HCO_3^- ions are
[o] _____

Buffer systems other than the carbonic acid–bicarbonate system release H^+ ions.

Combined Effects
[p] _____ P_{CO_2}
[q] _____ H^+
[r] _____ HCO_3

2. Section integration

After falling into an abandoned stone quarry filled with water and nearly drowning, a young boy is rescued. His rescuers assess his condition and find that his body fluids have high P_{CO_2} and lactate levels, and low P_{O_2} levels. Identify the underlying problem and recommend the necessary treatment to restore homeostatic conditions. _____

_____

Visual Outline with Key Terms

Summarize the content of each module using the terms in the order provided.

SECTION 1

Fluid and Electrolyte Balance

- minerals
- electrolytes
- intracellular fluid (ICF)
- extracellular fluid (ECF)
- fluid compartments

_____

_____

_____

_____

_____

_____

_____

_____

24.1

Fluid balance exists when water gains equal water losses

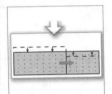

- fluid balance
- fluid shift
- dehydration

_____

_____

_____

_____

_____

_____

_____

24.2

Mineral balance involves balancing electrolyte gains and losses

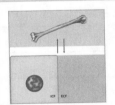

- mineral balance
- ion absorption
- ion excretion

_____

_____

_____

_____

_____

_____

• = Term boldfaced in this module

24.3

Water balance depends on sodium balance, and the two are regulated simultaneously

- sodium balance
- osmoreceptors
- hyponatremia
- hypernatremia

24.4

Disturbances of potassium balance are uncommon but extremely dangerous

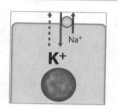

- potassium balance
- hypokalemia
- hyperkalemia
- aldosteronism

SECTION 2

Acid-Base Balance

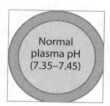

- acid-base balance
- fixed acids
- organic acids
- volatile acids

• = *Term boldfaced in this module*

24.5

Potentially dangerous disturbances in acid-base balance are opposed by buffer systems

- acidemia
- acidosis
- alkalemia
- alkalosis
- carbonic acid
- buffer system

24.6

Buffer systems can delay but not prevent pH shifts in the ICF and ECF

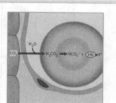

- phosphate buffer system
- protein buffer systems
- carbonic acid–bicarbonate buffer system
- hemoglobin buffer system
- bicarbonate reserve
- metabolic acid-base disorders
- respiratory acid-base disorders

24.7

The homeostatic responses to acidosis and alkalosis involve respiratory and renal mechanisms as well as buffer systems

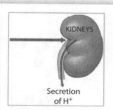

- metabolic acidosis
- carbonic acid–bicarbonate buffer system
- respiratory response to acidosis
- renal response to acidosis
- metabolic alkalosis
- respiratory response to alkalosis
- renal response to alkalosis

= Term boldfaced in this module

Respiratory acid-base disorders are the most common challenges to acid-base balance

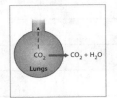

- respiratory acid-base disorders
- respiratory acidosis
- respiratory compensation
- renal compensation
- respiratory alkalosis

1. Labeling

Label the structures of the male reproductive system in the accompanying diagram.

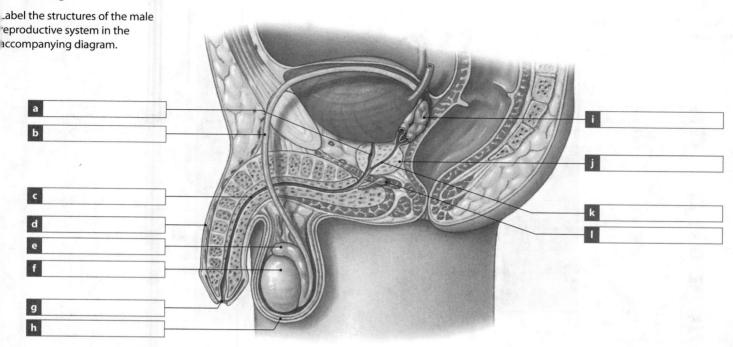

a _____
b _____
c _____
d _____
e _____
f _____
g _____
h _____

i _____
j _____
k _____
l _____

2. Matching

Match the following terms with the most closely related description.

semen

epididymis

nurse cells

corpus spongiosum

luteinizing hormone (LH)

impotence

follicle-stimulating hormone (FSH)

spermatogonia

seminiferous tubules

dartos muscle

spermatogenesis

interstitial cells

spermiogenesis

penis and scrotum

a _____ Scrotal smooth muscle

b _____ Sperm stem cells

c _____ Sites of sperm production

d _____ Produce testosterone

e _____ Physical maturation of spermatids

f _____ Sperm production

g _____ External genitalia

h _____ Start of male reproductive tract

i _____ Maintain blood–testis barrier

j _____ Spermatozoa, seminal gland, and other gland secretions

k _____ Inability to achieve or maintain an erection

l _____ Erectile tissue surrounding the urethra

m _____ Induces secretion of androgens

n _____ Hormone that targets nurse cells

3. Section integration

In males, the endocrine disorder hypogonadism is primarily due to the underproduction of testosterone or the lack of tissue sensitivity to testosterone, and results in sterility. What are five primary functions of testosterone in males? _____

_____

_____

1. Labeling

Label the structures of the male reproductive system in the accompanying diagram.

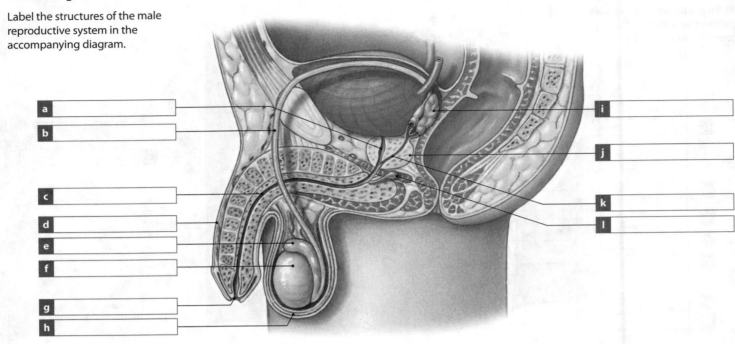

a _____

b _____

c _____

d _____

e _____

f _____

g _____

h _____

i _____

j _____

k _____

l _____

2. Matching

Match the following terms with the most closely related description.

- semen
- epididymis
- nurse cells
- corpus spongiosum
- luteinizing hormone (LH)
- impotence
- follicle-stimulating hormone (FSH)
- spermatogonia
- seminiferous tubules
- dartos muscle
- spermatogenesis
- interstitial cells
- spermiogenesis
- penis and scrotum

a _____ Scrotal smooth muscle

b _____ Sperm stem cells

c _____ Sites of sperm production

d _____ Produce testosterone

e _____ Physical maturation of spermatids

f _____ Sperm production

g _____ External genitalia

h _____ Start of male reproductive tract

i _____ Maintain blood–testis barrier

j _____ Spermatozoa, seminal gland, and other gland secretions

k _____ Inability to achieve or maintain an erection

l _____ Erectile tissue surrounding the urethra

m _____ Induces secretion of androgens

n _____ Hormone that targets nurse cells

3. Section integration

In males, the endocrine disorder hypogonadism is primarily due to the underproduction of testosterone or the lack of tissue sensitivity to testosterone, and results in sterility. What are five primary functions of testosterone in males? _____

_____

_____

1. Labeling

Label the structures of the female reproductive system in the accompanying diagram.

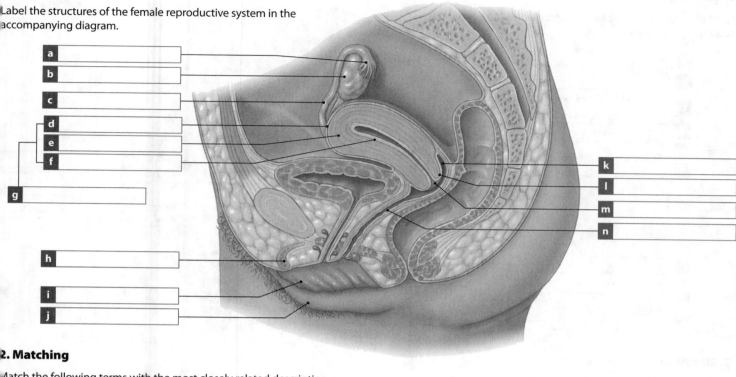

a _____
b _____
c _____
d _____
e _____
f _____
g _____
h _____
i _____
j _____

k _____
l _____
m _____
n _____

2. Matching

Match the following terms with the most closely related description.

LH surge

rectouterine pouch

tubal ligation

menarche

ovaries

corpus luteum

vulva

cervix

broad ligament

oocytes

GnRH

vesicouterine pouch

lactation

uterine cycle

a _____ Immature female gametes

b _____ Puberty in female

c _____ Pocket anterior to the uterus

d _____ Encloses the ovaries, uterine tubes, and uterus

e _____ Pocket posterior to the uterus

f _____ Endocrine structure

g _____ Averages 28 days

h _____ Milk production

i _____ Oocyte and hormone production

j _____ Triggers ovulation

k _____ Inferior portion of the uterus

l _____ Female surgical sterilization

m _____ Stimulates FSH production and secretion

n _____ Contains female external genitalia

3. Section integration

In a condition known as endometriosis, endometrial cells are believed to migrate from the body of the uterus either into the uterine tubes or through the uterine tubes and into the peritoneal cavity, where they become established. Explain why periodic pain is a major symptom of endometriosis. _____

_____

_____

1. Labeling

Label the structures of the female reproductive system in the accompanying diagram.

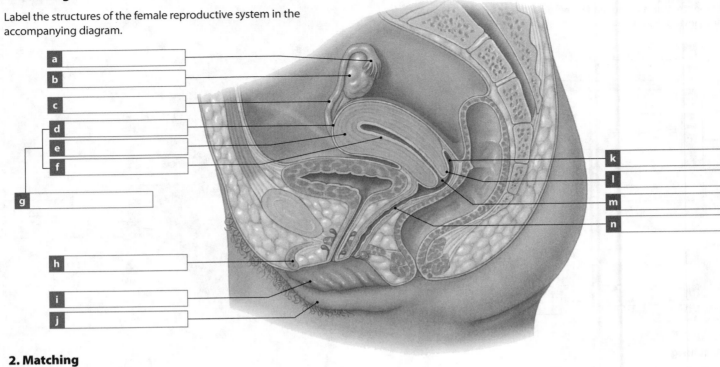

a _____

b _____

c _____

d _____

e _____

f _____

g _____

h _____

i _____

j _____

k _____

l _____

m _____

n _____

2. Matching

Match the following terms with the most closely related description.

- LH surge
- rectouterine pouch
- tubal ligation
- menarche
- ovaries
- corpus luteum
- vulva
- cervix
- broad ligament
- oocytes
- GnRH
- vesicouterine pouch
- lactation
- uterine cycle

a _____ Immature female gametes

b _____ Puberty in female

c _____ Pocket anterior to the uterus

d _____ Encloses the ovaries, uterine tubes, and uterus

e _____ Pocket posterior to the uterus

f _____ Endocrine structure

g _____ Averages 28 days

h _____ Milk production

i _____ Oocyte and hormone production

j _____ Triggers ovulation

k _____ Inferior portion of the uterus

l _____ Female surgical sterilization

m _____ Stimulates FSH production and secretion

n _____ Contains female external genitalia

3. Section integration

In a condition known as endometriosis, endometrial cells are believed to migrate from the body of the uterus either into the uterine tubes or through the uterine tubes and into the peritoneal cavity, where they become established. Explain why periodic pain is a major symptom of endometriosis. _____

_____

_____

Visual Outline with Key Terms

Summarize the content of each module using the terms in the order provided.

SECTION 1

The Male Reproductive System

- gonads
- external genitalia
- testis
- spermatozoa
- male reproductive tract
- semen

○ accessory organs
- ductus deferens
- seminal glands
- prostate gland
- urethra
- epididymis
○ external genitalia

- penis
- scrotum

_____
_____
_____
_____
_____
_____
_____
_____
_____
_____

25.1

The coiled seminiferous tubules of the testes are connected to the male reproductive tract

- epididymis
- ductus deferens
- ejaculatory duct
- seminal glands
- prostate gland
- bulbo-urethral glands
- scrotum
- penis

- superficial inguinal ring
- cremaster muscle
- dartos muscle
- inguinal canal
- inguinal hernias
- spermatic cords
- scrotal cavities
- scrotal septum

- raphe
- tunica albuginea
- seminiferous tubule
- rete testis
- efferent ductules

_____
_____
_____
_____
_____
_____
_____
_____
_____
_____
_____
_____
_____

25.2

Meiosis in the testes produces haploid spermatids that mature into spermatozoa

- spermatogenesis
- mitosis
- meiosis
- spermiogenesis
- diploid
- meiosis I
- meiosis II
- haploid
- synapsis

- tetrad
- spermatogonia
- primary spermatocyte
- secondary spermatocytes
- spermatids
- spermatozoon
- acrosomal cap

- head
- neck
- middle piece
- tail
- flagellum

_____
_____
_____
_____
_____
_____
_____
_____
_____
_____
_____

Term boldfaced in this module

25.3

Meiosis and early spermiogenesis occur within the seminiferous tubules

- ○ seminiferous tubules
- interstitial cells
- nurse cells
- spermiation
- blood–testis barrier
- basal compartment
- luminal compartment

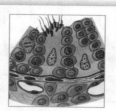

_____

_____

25.4

The male reproductive tract receives secretions from the seminal, prostate, and bulbo-urethral glands

- capacitation
- ampulla
- ejaculatory duct
- epididymis
- head (of epididymis)
- body (of epididymis)
- tail (of epididymis)
- stereocilia
- ductus deferens
- seminal glands
- semen
- prostate gland
- seminalplasmin
- bulbo-urethral glands

_____

_____

25.5

The penis conducts semen and urine to the exterior

- penis
- root
- body
- glans
- neck
- erectile tissue
- corpora cavernosa
- corpus spongiosum
- prepuce
- smegma
- arousal
- nitric oxide
- erection
- emission
- semen
- ejaculation
- male orgasm
- impotence

_____

_____

● = _Term boldfaced in this module_

25.6

Testosterone plays a key role in the establishment and maintenance of male sexual function

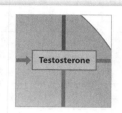

- luteinizing hormone (LH)
- follicle-stimulating hormone (FSH)
- ○ testosterone
- androgen-binding protein (ABP)
- ○ inhibin
- dihydrotestosterone (DHT)

_____

_____

_____

_____

_____

_____

_____

_____

SECTION 2

The Female Reproductive System

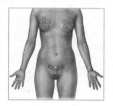

- oocytes
- ova
- female reproductive tract
- uterine tubes
- uterus
- vagina
- mammary glands
- ovaries
- ○ external genitalia
- clitoris
- labia

_____

_____

_____

_____

_____

_____

_____

_____

_____

25.7

The ovaries and the female reproductive tract are in close proximity but are not directly connected

- ovaries
- oocytes
- estrogens
- progestins
- inhibin
- uterine tube
- infundibulum
- uterus
- vesicouterine pouch
- clitoris
- labia
- rectouterine pouch
- vagina
- ovarian ligament
- mesovarium
- broad ligament
- suspensory ligament
- cortex (of ovary)
- medulla (of ovary)
- germinal epithelium
- tunica albuginea

_____

_____

_____

_____

_____

_____

_____

_____

_____

_____

_____

= _Term boldfaced in this module_

25.8

Meiosis I in the ovaries produces a single haploid secondary oocyte that completes meiosis II only if fertilization occurs

- oogenesis
- primary oocyte
- ovum
- polar bodies
- secondary oocytes
- oogonia
- ovarian follicles
- egg nests
- primordial follicle

- granulosa cells
- zona pellucida
- thecal cells
- secondary follicles
- tertiary follicle
- antrum
- corona radiata
- ovulation
- follicular phase

- luteal phase
- corpus luteum
- corpus albicans
- atresia

25.9

The uterine tubes are connected to the uterus, a hollow organ with thick muscular walls

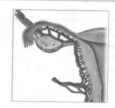

- uterine tube
- uterus
- infundibulum
- fimbriae
- ampulla (of the uterine tube)
- isthmus (of the uterine tube)
- perimetrium

- myometrium
- endometrium
- uterine cavity
- internal os
- cervical canal
- external os
- embryo
- fetus
- fundus (of the uterus)

- body (of the uterus)
- cervix (of the uterus)

25.10

The uterine cycle involves changes in the functional zone of the endometrium

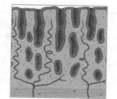

- arcuate arteries
- radial arteries
- basilar zone
- functional zone
- straight arteries
- spiral arteries
- uterine cycle

- menses
- menstruation
- proliferative phase
- secretory phase
- menstrual cycle
- menarche
- menopause

• = *Term boldfaced in this module*

25.11

The entrance to the vagina is enclosed by external genitalia

- vagina
- vestibule
- vaginal canal
- fornix
- rugae
- hymen
- vulva
- pudendum
- labia minora
- lesser vestibular glands
- greater vestibular glands
- mons pubis
- prepuce
- clitoris
- labia majora

25.12

The mammary glands nourish the infant after delivery

- mammary glands
- lactation
- pectoral fat pad
- suspensory ligaments of the breast
- lobes
- lobules
- secretory alveoli
- lactiferous duct
- lactiferous sinus
- nipple
- areola

25.13

The ovarian and uterine cycles are regulated by hormones of the hypothalamus, pituitary gland, and ovaries

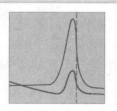

- ovarian cycle
- uterine cycle
- follicular phase
- estrogens
- estradiol
- luteal phase
- progesterone
- basal body temperature

= *Term boldfaced in this module*

25.14

Birth control strategies vary in effectiveness and in the nature of associated risks

- male condoms
- sexually transmitted diseases (STDs)
- diaphragm
- oral contraceptives
- progesterone-only pill
- Depo-Provera
- intrauterine device (IUD)
- rhythm method
- hormonal post-coital contraception
- vasectomy
- tubal ligation

_____

_____

_____

_____

_____

_____

_____

_____

_____

_____

25.15

Reproductive system disorders are relatively common and often deadly

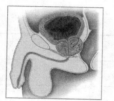

- benign prostatic hypertrophy (BPH)
- prostate cancer
- prostate-specific antigen (PSA)
- prostatectomy
- testicular cancer
- cysts
- fibrocystic disease
- breast cancer
- ovarian cancer
- carcinomas
- cervical cancer
- human papillomaviruses (HPV)

_____

_____

_____

_____

_____

_____

_____

_____

_____

_____

• = _Term boldfaced in this module_

1. Matching

Match the following terms with the most closely related description.

- hCG
- conception
- chorion
- syncytial trophoblast
- colostrum
- amnion
- amphimixis
- morula
- embryonic disc
- neonate
- gestation
- inner cell mass
- relaxin
- blastocyst

a	_____	Fertilization
b	_____	Newborn infant
c	_____	Pronuclei fuse
d	_____	Period of prenatal development
e	_____	Pregnancy test
f	_____	Softens pubic symphysis
g	_____	Mammary gland secretion
h	_____	Mesoderm and ectoderm
i	_____	Hollow ball of cells
j	_____	Mesoderm and trophoblast
k	_____	Forms the embryo
l	_____	Cytoplasm with many nuclei
m	_____	Solid ball of cells
n	_____	Gastrulation product

2. Multiple choice

Choose the bulleted item that best completes each statement.

a Fertilization typically occurs in the _____ .
- lower part of the uterine tube
- upper part of the uterus
- junction between the ampulla and isthmus of the uterine tube
- cervix

b Fetal development begins at the start of the
_____ .
- implantation process
- second month after fertilization
- ninth week after fertilization
- sixth month after fertilization

c Organs and organ systems complete most of their development by the end of the _____ .
- first trimester
- second trimester
- third trimester
- expulsion stage

d The four general processes that occur during the first trimester include _____ .
- blastomere, blastocyst, morula, and trophoblast
- cleavage, implantation, placentation, and embryogenesis
- placentation, dilation, expulsion, and organogenesis
- yolk sac, amnion, allantois, and chorion

e The most dangerous period in prenatal or neonatal life is the _____ .
- first trimester
- second trimester
- third trimester
- expulsion stage

f The systems that were relatively nonfunctional during the fetal period that must become functional at birth are the _____ systems.
- cardiovascular, muscular, and skeletal
- integumentary, reproductive, and nervous
- respiratory, digestive, and urinary
- endocrine, nervous, and digestive

3. Section integration

Tina gives birth to a baby with a congenital deformity of the stomach. Tina believes that her baby's affliction is the result of a viral infection that she suffered during her third trimester. Is this a possibility? Explain. _____

_____

_____

1. Matching

Match the following terms with the most closely related description.

- hCG
- conception
- chorion
- syncytial trophoblast
- colostrum
- amnion
- amphimixis
- morula
- embryonic disc
- neonate
- gestation
- inner cell mass
- relaxin
- blastocyst

a	_____	Fertilization
b	_____	Newborn infant
c	_____	Pronuclei fuse
d	_____	Period of prenatal development
e	_____	Pregnancy test
f	_____	Softens pubic symphysis
g	_____	Mammary gland secretion
h	_____	Mesoderm and ectoderm
i	_____	Hollow ball of cells
j	_____	Mesoderm and trophoblast
k	_____	Forms the embryo
l	_____	Cytoplasm with many nuclei
m	_____	Solid ball of cells
n	_____	Gastrulation product

2. Multiple choice

Choose the bulleted item that best completes each statement.

a Fertilization typically occurs in the _____.
- lower part of the uterine tube
- upper part of the uterus
- junction between the ampulla and isthmus of the uterine tube
- cervix

b Fetal development begins at the start of the _____.
- implantation process
- second month after fertilization
- ninth week after fertilization
- sixth month after fertilization

c Organs and organ systems complete most of their development by the end of the _____.
- first trimester
- second trimester
- third trimester
- expulsion stage

d The four general processes that occur during the first trimester include _____.
- blastomere, blastocyst, morula, and trophoblast
- cleavage, implantation, placentation, and embryogenesis
- placentation, dilation, expulsion, and organogenesis
- yolk sac, amnion, allantois, and chorion

e The most dangerous period in prenatal or neonatal life is the _____.
- first trimester
- second trimester
- third trimester
- expulsion stage

f The systems that were relatively nonfunctional during the fetal period that must become functional at birth are the _____ systems.
- cardiovascular, muscular, and skeletal
- integumentary, reproductive, and nervous
- respiratory, digestive, and urinary
- endocrine, nervous, and digestive

3. Section integration

Tina gives birth to a baby with a congenital deformity of the stomach. Tina believes that her baby's affliction is the result of a viral infection that she suffered during her third trimester. Is this a possibility? Explain. _____

_____

_____

1. Matching

Match the following terms with the most closely related description.

- genotype
- heterozygous
- locus
- autosomes
- simple inheritance
- homozygous
- alleles
- polygenic inheritance
- homologous
- genetics
- karyotype
- phenotype

a _____ Visible characteristics

b _____ Alternate forms of a gene

c _____ Refers to the two members of a pair of chromosomes

d _____ Array of the entire set of chromosomes in a cell

e _____ An individual's chromosomes and genes

f _____ Gene's position on a chromosome

g _____ Two different alleles for the same gene

h _____ Study of the mechanisms of inheritance

i _____ Two identical alleles for the same gene

j _____ Interactions between alleles on several genes

k _____ Phenotype determined by a single pair of alleles

l _____ Chromosomes affecting somatic characteristics

2. Section integration

Use the Punnett squares below to answer the questions concerning the following genetic conditions.

a Tongue rolling is inherited as a dominant trait (T). Explain how it is possible for two parents who are tongue rollers to have children who do not have the ability to roll the tongue.

b Achondroplasia dwarfism is an autosomal genetic disorder that results from problems with the replacement of cartilage by bone in the arms and legs. Assume that two dwarf parents have a normal-sized child. Determine the genotype of the parents and predict the probability that a second child would also be normal sized.

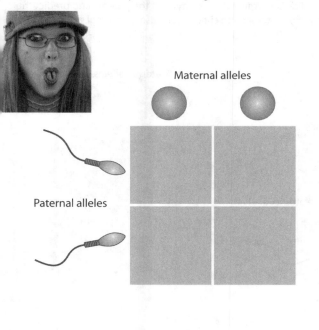

Maternal alleles

Paternal alleles

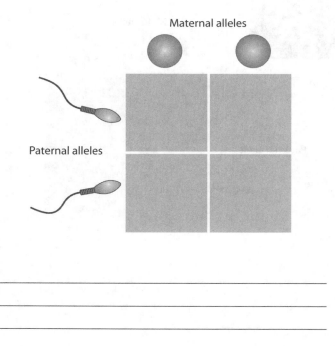

Maternal alleles

Paternal alleles

_____

_____

_____

_____

_____

_____

1. Matching

Match the following terms with the most closely related description.

- genotype
- heterozygous
- locus
- autosomes
- simple inheritance
- homozygous
- alleles
- polygenic inheritance
- homologous
- genetics
- karyotype
- phenotype

a _____ Visible characteristics

b _____ Alternate forms of a gene

c _____ Refers to the two members of a pair of chromosomes

d _____ Array of the entire set of chromosomes in a cell

e _____ An individual's chromosomes and genes

f _____ Gene's position on a chromosome

g _____ Two different alleles for the same gene

h _____ Study of the mechanisms of inheritance

i _____ Two identical alleles for the same gene

j _____ Interactions between alleles on several genes

k _____ Phenotype determined by a single pair of alleles

l _____ Chromosomes affecting somatic characteristics

2. Section integration

Use the Punnett squares below to answer the questions concerning the following genetic conditions.

a Tongue rolling is inherited as a dominant trait (T). Explain how it is possible for two parents who are tongue rollers to have children who do not have the ability to roll the tongue.

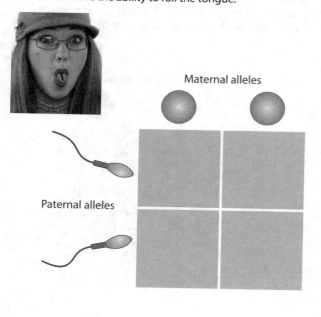

Maternal alleles

Paternal alleles

_____
_____
_____

b Achondroplasia dwarfism is an autosomal genetic disorder that results from problems with the replacement of cartilage by bone in the arms and legs. Assume that two dwarf parents have a normal-sized child. Determine the genotype of the parents and predict the probability that a second child would also be normal sized.

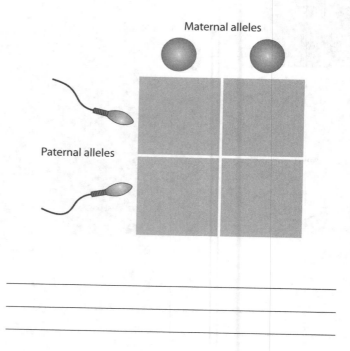

Maternal alleles

Paternal alleles

_____
_____
_____

Visual Outline with Key Terms

Summarize the content of each module using the terms in the order provided.

An Overview of Development

- development
- embryological development
- embryology
- fetal development
- prenatal development
- gestation
- first trimester
- second trimester
- third trimester
- postnatal development
- maturity

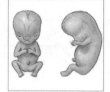

_____
_____
_____
_____
_____
_____
_____
_____
_____

26.1

At fertilization, a secondary oocyte and a spermatozoon form a zygote that prepares for cell division

- fertilization
- zygote
- oocyte activation
- female pronucleus
- male pronucleus
- cleavage
- amphimixis
- blastomeres

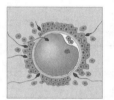

_____
_____
_____
_____
_____
_____
_____
_____

26.2

Cleavage continues until the blastocyst implants in the uterine wall

- pre-embryo
- morula
- blastocyst
- blastocoele
- implantation
- trophoblast
- inner cell mass
- cellular trophoblast
- syncytial trophoblast
- lacunae
- villi
- amniotic cavity

_____
_____
_____
_____
_____
_____
_____
_____

● = Term boldfaced in this module

26.3

Gastrulation produces three germ layers: ectoderm, endoderm, and mesoderm

- blastodisc
- amnion
- extra-embryonic membranes
- yolk sac
- primitive streak
- ectoderm
- endoderm
- mesoderm
- germ layers
- gastrulation
- embryonic disc
- gestational trophoblastic neoplasia

26.4

The extra-embryonic membranes form the placenta that supports fetal growth and development

- yolk sac
- amnion
- amniotic fluid
- allantois
- chorion
- placenta

26.5

The formation of extra-embryonic membranes is associated with major changes in the shape and complexity of the embryo

- head fold
- chorionic villi
- tail fold
- body stalk
- yolk stalk
- umbilical stalk
- umbilical cord

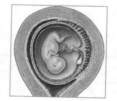

• = *Term boldfaced in this module*

26.6

The placenta performs many vital functions for the duration of prenatal development

- umbilical arteries
- umbilical vein
- human chorionic gonadotropin (hCG)
- human placental lactogen (hPL)
- relaxin
- progesterone
- estrogens

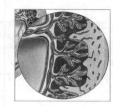

26.7

Organ systems are established in the first trimester and become functional in the second and third trimesters

- organogenesis
- neural plate

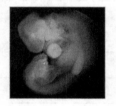

26.8

Pregnancy places anatomical and physiological stresses on maternal systems

- pregnancy
- anatomical changes
- physiological stresses

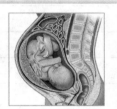

• = *Term boldfaced in this module*

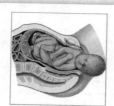

26.9

Multiple factors initiate and accelerate the process of labor

- false labor
- true labor
- positive feedback
- parturition
- dilation stage
- expulsion stage
- delivery
- placental stage
- afterbirth
- premature labor
- premature delivery

26.10

After delivery, development initially requires nourishment by maternal systems

- neonatal period
- milk let-down reflex
- colostrum
- lysozyme
- oxytocin
- life stages
- puberty

26.11

At puberty, male and female sex hormones have differential effects on most body systems

○ responses to testosterone in males
○ responses to estrogens in females

• = *Term boldfaced in this module*

Genetics and Inheritance

- inheritance
- genetics
- genotype
- phenotype
- karyotype
- homologous
  chromosomes
- autosomal
  chromosomes
- sex chromosomes
- X chromosome
- Y chromosome

26.12

Genes and chromosomes determine patterns of inheritance

- locus
- alleles
- homozygous
- heterozygous
- simple inheritance
- strict dominance
- recessive
- Punnett squares
- polygenic
  inheritance

26.13

There are several different patterns of inheritance

- codominance
- X-linked
- ○ carrier

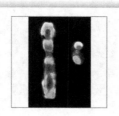

● = *Term boldfaced in this module*

26.14

Thousands of clinical disorders have been linked to abnormal chromosomes and/or genes

- single nucleotide polymorphisms (SNPs)
- trisomy 21
- Klinefelter syndrome
- Turner syndrome
- monosomy
- human genome

_____

_____

_____

_____

_____

_____

_____

_____

_____

• = *Term boldfaced in this module*

Chapter 1

Section 1 Review

1. **a.** respiration; **b.** growth and reproduction; **c.** adaptability; **d.** circulation; **e.** excretion; **f.** digestion; **g.** movement; **h.** responsiveness

2. Anatomy: right atrium, myocardium, left ventricle, endocardium, superior vena cava; Physiology: valve to aorta opens, valve between left atrium and left ventricle closes, pressure in left atrium, electrocardiogram

3. Both keys and messenger molecules have specific three-dimensional shapes, so both can "fit" into and function with their complementary structures (a lock or a receptor protein) only if the shapes match closely enough. Because messenger molecules need not be flat in one dimension (as keys tend to be), they can be very complex in shape, and thus may be able to bind with a variety of complementary receptor proteins. Moreover, though a lock has moving parts, it does not actually change shape, whereas a receptor protein bound by a messenger can change shape, which in turn affects its function.

4. Larger organisms simply cannot absorb the amount of needed materials or excrete wastes rapidly enough across their body surfaces as very small organisms can. Their food must be processed, or digested, before being absorbed, and because the processes of absorption, respiration, and excretion occur in different parts of the organism, an efficient, internal circulation network for their materials is necessary for survival.

Section 2 Review

1. cardiovascular system: blood cells; digestive system: smooth muscle cell; reproductive system: sperm cell or oocyte; skeletal system: bone cell (osteocyte); nervous system: nerve cell (neuron). (Other cells could be listed for other systems.)

2. **a.** cells; **b.** organs; **c.** organ systems; **d.** epithelial tissue; **e.** external and internal surfaces; **f.** glandular secretions; **g.** connective tissue; **h.** matrix;
i. protein fibers; **j.** ground substance; **k.** muscle tissue; **l.** movement; **m.** bones of the skeleton; **n.** blood; **o.** materials within digestive tract; **p.** nervous tissue; **q.** neuroglia

3. tissue: c; cell: b; organ: d; molecule: a; organism: f; organ system: e

4. **a.** skeletal system: support, protection of soft tissues, mineral storage, blood formation; **b.** digestive system: processing of food and absorption of nutrients, minerals, vitamins, and water; **c.** integumentary system: protection from environmental hazards, temperature control; **d.** urinary system: elimination of excess water, salts, and waste products, and control of pH; **e.** nervous system: directing immediate responses to stimuli, usually by coordinating the activities of other organ systems

Section 3 Review

1. **a.** positive feedback; **b.** homeostatic regulation; **c.** homeostasis; **d.** negative feedback; **e.** negative feedback

2. **a.** negative feedback; **b.** positive feedback; **c.** negative feedback; **d.** positive feedback

3. **a.** blood flow to skin increases, sweating increases, body surface cools, temperature declines; **b.** blood flow to skin decreases, shivering occurs, body heat is conserved, temperature rises

4. One reason your body temperature may have dropped is that your body may be losing heat faster than it is being produced. (This is more likely to occur on a cool day.) Perhaps hormones have caused a decrease in your metabolic rate, so your body is not producing as much heat as it normally would. Or you may have an infection that has temporarily reset the set point of the body's "thermostat" to a value higher than normal. The last possibility is the most likely explanation given the circumstances.

Section 4 Review

1. **a.** superior; **b.** inferior; **c.** cranial; **d.** caudal; **e.** posterior or dorsal; **f.** anterior or ventral; **g.** lateral; **h.** medial; **i.** proximal; **j.** distal; **k.** proximal; **l.** distal

2. **a.** thoracic cavity; **b.** mediastinum; **c.** left lung; **d.** trachea, esophagus; **e.** heart; **f.** diaphragm; **g.** abdominopelvic cavity; **h.** peritoneal cavity; **i.** digestive glands and organs; **j.** pelvic cavity; **k.** reproductive organs

Chapter 2

Section 1 Review

1. **a.** 2 **b.** 4 **c.** 1 **d.** 0 **e.** 6 **f.** 12 **g.** 7 **h.** 7 **i.** 20 **j.** 20

2. **a.** element; **b.** compound; **c.** element; **d.** compound

3. **a.** ions; **b.** neutrons; **c.** compound; **d.** atomic number; **e.** hydrogen bond; **f.** covalent bond; **g.** mass number; **h.** element; **i.** protons; **j.** isotopes; **k.** ionic bond; **l.** electrons

4. **a.** Each element consists of atoms containing a characteristic number of protons (atomic number). However, the outer energy level of inert elements is filled with electrons, whereas the outer energy level of reactive elements is not filled with electrons.

   b. Both polar and nonpolar molecules are held together by covalent bonds. However, in a polar molecule the electrons are not shared equally, so it carries small positive and negative charges on its surface; in a nonpolar molecule the electrons are shared equally, so it is electrically neutral.

   c. Both covalent bonds and ionic bonds bind atoms together, but covalent bonds involve the sharing of electrons between atoms, whereas ionic bonds involve the electrical attraction of oppositely charged atoms (ions).

Section 2 Review

1. **a.** H_2 **b.** $2H$ **c.** $6H_2O$ **d.** $C_{12}H_{22}O_{11}$

2. $C_6H_{12}O_6 + 6O_2 \rightarrow 6CO_2 + 6H_2O$

3. **a.** hydrolysis reaction **b.** dehydration synthesis

4. **a.** enzyme; **b.** reactants; **c.** hydrolysis; **d.** endergonic; **e.** exchange reaction; **f.** organic compounds; **g.** exergonic; **h.** activation energy

5. A decreased amount of enzyme at the second step would limit the amount of the intermediate products in the next two steps. This would cause a decrease in the amount of the final product.

Section 3 Review

1. Four important properties of water in the human body are effective lubrication (as between bony surfaces in a joint), reactivity (participates in chemical reactions), high heat capacity (readily absorbs and retains heat), and solubility (is a solvent for many substances).

2. **a.** inorganic compounds; **b.** solute; **c.** alkaline; **d.** hydrophilic; **e.** solvent; **f.** salt; **g.** buffers; **h.** acid; **i.** hydrophobic; **j.** water

3. **a.** acidic; **b.** neutral; **c.** alkaline; **d.** The pH 3 solution is 1000 times more acidic than the pH 6 solution—it contains a thousand-fold (10^3) increase in the concentration of hydrogen ions (H^+).

   e. Three negative effects of abnormal fluctuations in pH are cell and tissue damage (due to broken bonds), changes in the shapes of proteins, and altered cellular functions.

4. Table salt dissociates or dissolves in pure water but since it does not release either hydrogen (H^+) ions or (OH^-) ions, no change in pH occurs.

Section 4 Review

1. **a.** glycogen; **b.** isomers; **c.** polyunsaturated; **d.** active site; **e.** glycerol; **f.** cholesterol; **g.** ATP; **h.** RNA; **i.** peptide bond; **j.** nucleotide; **k.** monosaccharide

2. **a.** polysaccharide, polyunsaturated, polypeptide; **b.** triglyceride; **c.** disaccharide, diglyceride, dipeptide; **d.** glycogen, glycolipids

3. **a.** carbohydrates; **b.** polysaccharides; **c.** disaccharides; **d.** monosaccharides; **e.** lipids; **f.** fatty acids; **g.** glycerol; **h.** proteins; **i.** amino acids; **j.** nucleic acids; **k.** RNA; **l.** DNA; **m.** nucleotides; **n.** ATP; **o.** phosphate groups

Chapter 3

Section 1 Review

1. **a.** microvilli: increased surface area to facilitate absorption of extracellular materials; **b.** Golgi apparatus: storage, alteration, and packaging of newly synthesized proteins; **c.** lysosome: intracellular removal of damaged organelles or of pathogens; **d.** mitochondrion:

production of 95 percent of the ATP required by the cell; **e.** peroxisome: neutralization of toxic compounds; **f.** nucleus: control of metabolism, storage and processing of genetic information, and control of protein synthesis; **g.** endoplasmic reticulum: synthesis of secretory products, and intracellular storage and transport; **h.** ribosomes: protein synthesis; **i.** cytoskeleton: provides strength and support, enables movement of cellular structures and materials

2. **a.** glycolysis; **b.** aerobic metabolism; **c.** microvilli; **d.** lysosome

3. Similar to the role of a plasma membrane around a cell, an organelle membrane physically isolates the organelle's contents from the cytosol, regulates exchange with the cytosol, and provides structural support.

Section 2 Review

1. **a.** chromosomes; **b.** thymine; **c.** nuclear envelope; **d.** introns; **e.** uracil; **f.** nucleoli; **g.** nuclear pore; **h.** transcription; **i.** gene; **j.** exons; **k.** mRNA; **l.** tRNA; **m.** nucleus; **n.** genetic information

2. **a.** AUG/UUU/UGU/GCC/GCC/UUA **b.** UAC/AAA/ACA/CGG/CGG/AAU **c.** Methionine-Phenylalanine-Cysteine-Alanine-Alanine-Leucine

3. The nucleus contains the information for synthesizing proteins in the nucleotide sequence of its DNA. Changes in the extracellular fluid can affect cells through the binding of molecules to plasma membrane receptors or by the diffusion of molecules through membrane channels. Such stimuli may result in alterations of genetic activity in the nucleus. These alterations may change biochemical processes and metabolic pathways through the synthesis of additional, fewer, or different enzymes. Altered genetic activity may also change the physical structure of the cell by synthesizing additional, fewer, or different structural proteins.

Section 3 Review

1. **a.** diffusion; **b.** facilitated diffusion; **c.** molecular size; **d.** net diffusion of water; **e.** active transport; **f.** specificity;

g. vesicular transport; **h.** exocytosis; **i.** pinocytosis; **j.** "cell eating"

2. **a.** diffusion; **b.** diffusion; **c.** neither; **d.** diffusion; **e.** osmosis; **f.** osmosis

Section 4 Review

1. **a.** somatic cells; **b.** G_1 phase; **c.** G_2 phase; **d.** DNA replication; **e.** mitosis; **f.** metaphase; **g.** telophase; **h.** cytokinesis

2. **a.** telophase; **b.** prophase; **c.** centromere

3. A cell that undergoes repeated rounds of the cell cycle without cytokinesis could result in a large, multinucleated cell.

Chapter 4

Section 1 Review

1. **a.** simple squamous epithelium; **b.** simple cuboidal epithelium; **c.** simple columnar epithelium; **d.** stratified squamous epithelium; **e.** stratified cuboidal epithelium; **f.** stratified columnar epithelium

2. **a.** pseudostratified columnar epithelium; **b.** urinary bladder, ureters, urine-collecting chambers in the kidney; **c.** stratified squamous epithelium; **d.** simple columnar epithelium; **e.** lining of the peritoneum and pericardium, exchange surfaces (alveoli) within the lungs; **f.** lining of exocrine glands and ducts, kidney tubules; **g.** stratified cuboidal epithelium

3. **a.** mucous cells; **b.** mucin; **c.** mucus; **d.** ducts; **e.** epithelial surfaces; **f.** exocrine glands; **g.** merocrine secretion; **h.** apocrine secretion; **i.** holocrine secretion; **j.** interstitial fluid; **k.** endocrine glands

4. **a.** avascular; **b.** alveolar (acinar) gland; **c.** transitional epithelium; **d.** occluding junction; **e.** basal lamina; **f.** simple gland; **g.** mesothelium (simple squamous epithelium)

Section 2 Review

1. **a.** loose connective tissue; **b.** adipose; **c.** regular; **d.** tendons; **e.** ligaments; **f.** fluid connective tissue; **g.** blood; **h.** hyaline; **i.** chondrocytes in lacunae; **j.** bone

2. **a.** perichondrium, periosteum, peritoneum, pericardium; **b.** osteocyte, periosteum;

c. chondrocyte, perichondrium; **d.** interstitial growth; **e.** lacunae; **f.** chondrocyte; **g.** osseous tissue; **h.** fibrous cartilage; **i.** adipocytes; **j.** synovial membrane; **k.** cutaneous membrane; **l.** perichondrium

Section 3 Review

1. **a.** cardiac; **b.** smooth; **c.** nonstriated; **d.** multinucleate

2. **a.** cell body; **b.** dendrites; **c.** neuroglia; **d.** maintain physical structure of neural tissue; **e.** repair neural tissue framework after injury; **f.** perform phagocytosis; **g.** provide nutrients to neurons; **h.** regulate the composition of interstitial fluid surrounding neurons

3. **a.** axon; **b.** intercalated disc; **c.** neuroglia; **d.** skeletal muscle tissue; **e.** smooth muscle tissue; **f.** regeneration; **g.** inflammation

4. Increased blood flow and blood vessel permeability enhance the delivery of oxygen and nutrients and the migration of additional phagocytes into the area, and the removal of toxins and waste products from the area.

Chapter 5

Section 1 Review

1. **a.** dermis: the connective tissue layer beneath the epidermis; **b.** epidermis: the protective epithelium covering the surface of the skin; **c.** papillary layer: vascularized areolar tissue containing capillaries, lymphatic vessels, and sensory neurons that supply the skin surface; **d.** reticular layer: interwoven meshwork of dense irregular connective tissue containing collagen fibers and elastic fibers; **e.** hypodermis (subcutaneous layer or superficial fascia): layer of loose connective tissue below the dermis

2. **a.** accessory structures; **b.** epidermis; **c.** granulosum; **d.** papillary layer; **e.** nerves; **f.** reticular layer; **g.** collagen; **h.** hypodermis; **i.** connective; **j.** fat

3a. Malignant melanoma is often fatal because melanocytes are located close to the dermal layer, so if they become malignant, they can easily metastasize

through the blood vessels and lymphatic vessels in nearby connective tissues.

3b. Fingers (and toes) swell up because of the hypotonic osmotic flow of water into the dead, keratinized cells of the outer layer of the epidermis, the stratum corneum. Because the underlying strata and dermis do not expand, the larger surface area of the swollen stratum corneum must go somewhere and it forms folds and creases, or wrinkles. Other areas of the body lack a thick stratum corneum, so little swelling and wrinkling result.

Section 2 Review

1. **a.** vitamin D_3; **b.** nail root **c.** apocrine sweat glands **d.** EGF **e.** reticular layer of dermis **f.** wrinkled skin **g.** sebum **h.** malignant melanoma **i.** merocrine sweat glands **j.** eponychium

2. **a.** free edge; **b.** lateral nail fold; **c.** nail body; **d.** lunula; **e.** proximal nail fold; **f.** eponychium; **g.** eponychium; **h.** proximal nail fold; **i.** nail root; **j.** lunula; **k.** nail body; **l.** hyponychium; **m.** phalanx; **n.** dermis; **o.** epidermis

3. **a.** hair shaft; **b.** sebaceous gland; **c.** arrector pili muscle; **d.** connective tissue sheath of hair bulb; **e.** root hair plexus

4. The chemicals in hair dyes break the protective covering of the cortex allowing the dyes to stain the medulla of the shaft. This is not permanent because the cortex remains damaged, allowing shampoo and UV rays from the sun to enter the medulla and affect the color. Also, the viable portion of the hair remains unaffected, so that when the shaft is replaced the color will be lost.

Chapter 6

Section 1 Review

1. **a.** irregular bones; **b.** epiphyses; **c.** fossa; **d.** medullary cavity; **e.** trabeculae; **f.** osteoclasts; **g.** sesamoid bones; **h.** osteogenesis; **i.** osteon; **j.** appositional growth; **k.** endochondral ossification

2. **a.** intramembranous ossification; **b.** collagen; **c.** osteocytes; **d.** lacunae; **e.** hyaline cartilage; **f.** periosteum; **g.** compact bone

. The fracture might have damaged the epiphyseal cartilage in Rebecca's right leg. Even though the bone healed properly, the damaged leg did not produce as much cartilage as did the undamaged leg. The result would be a shorter bone on the side of the injury.

Section 2 Review

. **a.** calcitonin; **b.** ↓ Ca^{2+} concentration in body fluids; **c.** ↓ calcium level; **d.** parathyroid glands; **e.** ↑ Ca^{2+} concentration in body fluids; **f.** releases stored Ca^{2+} from bone; **g.** homeostasis

. **a.** spiral fracture; **b.** transverse fracture; **c.** greenstick fracture; **d.** comminuted fracture; **e.** compression fracture; **f.** Colles fracture (typically results from cushioning a fall)

Chapter 7

Section 1 Review

. **a.** axial; **b.** longitudinal; **c.** skull; **d.** mandible; **e.** lacrimal; **f.** occipital; **g.** temporal; **h.** hyoid; **i.** vertebral column; **j.** thoracic; **k.** lumbar; **l.** sacral; **m.** floating; **n.** sternum; **o.** xiphoid process

. **a.** cervical; **b.** thoracic; **c.** lumbar

Section 2 Review

. **a.** clavicle; **b.** scapula; **c.** humerus; **d.** radius; **e.** ulna; **f.** carpal bones; **g.** metacarpal bones; **h.** phalanges; **i.** hip bone (coxal bone); **j.** femur; **k.** patella; **l.** tibia; **m.** fibula; **n.** tarsal bones; **o.** metatarsal bones; **p.** phalanges

. **a.** anterior view; **b.** lateral view; **c.** posterior view; **d.** acromion; **e.** coracoid process; **f.** scapular spine; **g.** glenoid cavity; **h.** subscapular fossa; **i.** supraspinous fossa; **j.** infraspinous fossa

. **a.** sacrum; **b.** coccyx; **c.** ilium; **d.** pubis; **e.** ischium; **f.** hip (coxal) bone; **g.** iliac fossa; **h.** acetabulum; **i.** obturator foramen

Chapter 8

Section 1 Review

. **a.** diarthrosis; **b.** pronation-supination; **c.** shoulder; **d.** articular discs;

e. synarthrosis; **f.** dislocation; **g.** fluid-filled pouch; **h.** amphiarthrosis

2. **a.** medullary cavity; **b.** spongy bone; **c.** periosteum; **d.** synovial membrane; **e.** articular cartilage; **f.** joint cavity (containing synovial fluid); **g.** joint capsule; **h.** compact bone

3. **a.** flexion; **b.** extension; **c.** hyperextension; **d.** flexion; **e.** hyperextension; **f.** abduction; **g.** adduction; **h.** head rotation; **i.** pronation; **j.** abduction; **k.** adduction; **l.** opposition

Section 2 Review

1. **a.** popliteal ligament; **b.** arthritis; **c.** osteoporosis; **d.** reinforce knee joint; **e.** disc outer layer; **f.** acetabulum; **g.** disc inner layer; **h.** dislocation

2. **a.** coracoclavicular ligaments; **b.** acromioclavicular ligament; **c.** tendon of supraspinatus muscle; **d.** acromion; **e.** articular capsule; **f.** subdeltoid bursa; **g.** synovial membrane; **h.** humerus; **i.** clavicle; **j.** coraco-acromial ligament; **k.** coracoid process; **l.** scapula; **m.** articular cartilages; **n.** joint cavity; **o.** glenoid labrum

3. **a.** patellar surface of femur; **b.** fibular collateral ligament; **c.** lateral condyle; **d.** lateral meniscus; **e.** tibia; **f.** fibula; **g.** posterior cruciate ligament (PCL); **h.** medial condyle; **i.** tibial collateral ligament; **j.** medial meniscus; **k.** anterior cruciate ligament (ACL)

4. The sternoclavicular joints are the only articulations between the pectoral girdles and the axial skeleton. The sacro-iliac joints are the articulations between the pelvic girdles and the axial skeleton.

Chapter 9

Section 1 Review

1. **a.** mitochondrion; **b.** sarcolemma; **c.** myofibril; **d.** thin filament; **e.** thick filament; **f.** triad; **g.** sarcoplasmic reticulum; **h.** T tubules; **i.** terminal cisterna; **j.** sarcoplasm; **k.** myofibril

2. **a.** I band; **b.** A band; **c.** H band; **d.** Z line; **e.** titin; **f.** zone of overlap; **g.** M line; **h.** thin filament; **i.** thick filament **j.** sarcomere

3. **a.** myoblast, myofibril, myofilament, myosatellite cell; **b.** sarcolemma, sarcoplasm, sarcomere, sarcoplasmic reticulum

Section 2 Review

1. **a.** isometric contraction; **b.** isotonic contraction; **c.** eccentric contraction; **d.** concentric contraction

2. **a.** fatty acids; **b.** O_2; **c.** glucose; **d.** glycogen; **e.** CP; **f.** creatine

3. **a.** Small; **b.** Intermediate; **c.** Large; **d.** Red; **e.** White; **f.** Low; **g.** Low; **h.** Dense; **i.** Intermediate; **j.** Many; **k.** Intermediate; **l.** Few; **m.** Prolonged; **n.** Rapid; **o.** Slow; **p.** Fast; **q.** Fast; **r.** Intermediate; **s.** Low; **t.** Low; **u.** High

Chapter 10

Section 1 Review

1. **a.** deltoid: multipennate; **b.** extensor digitorum: unipennate; **c.** rectus femoris: bipennate

2. **a.** sternocleidomastoid; **b.** deltoid; **c.** biceps brachii; **d.** external oblique; **e.** pronator teres; **f.** brachioradialis; **g.** flexor carpi radialis; **h.** rectus femoris; **i.** vastus lateralis; **j.** vastus medialis; **k.** gastrocnemius; **l.** soleus; **m.** pectoralis major; **n.** rectus abdominis; **o.** iliopsoas; **p.** tensor fasciae latae; **q.** gracilis; **r.** sartorius; **s.** tibialis anterior; **t.** extensor digitorum longus

Section 2 Review

1. **a.** occipitofrontalis (frontal belly); **b.** temporalis; **c.** orbicularis oculi; **d.** levator labii superioris; **e.** zygomaticus minor; **f.** zygomaticus major; **g.** buccinator; **h.** orbicularis oris; **i.** risorius; **j.** depressor labii inferioris; **k.** depressor anguli oris; **l.** masseter

2. **a.** mylohyoid; **b.** digastric; **c.** geniohyoid; **d.** omohyoid; **e.** stylohyoid; **f.** thyrohyoid; **g.** sternothyroid; **h.** sternohyoid; **i.** sternocleidomastoid

Section 3 Review

1. **a.** triceps brachii; **b.** anconeus; **c.** extensor carpi ulnaris; **d.** extensor carpi radialis brevis; **e.** extensor digitorum; **f.** flexor carpi ulnaris

2. **a.** gluteus medius; **b.** tensor fasciae latae; **c.** gluteus maximus; **d.** adductor magnus; **e.** gracilis; **f.** biceps femoris; **g.** semitendinosus; **h.** semimembranosus; **i.** sartorius; **j.** popliteus

3. **a.** gastrocnemius; **b.** tibialis anterior; **c.** fibularis longus; **d.** soleus; **e.** extensor digitorum longus; **f.** fibularis brevis; **g.** superior extensor retinaculum; **h.** calcaneal tendon; **i.** inferior extensor retinaculum

Chapter 11

Section 1 Review

1. **a.** neuron, neuroglia, neurofilaments, neurofibrils, neurotubules, neurilemma; **b.** dendrite, dendritic spines, telodendria, oligodendrocytes; **c.** efferent division, efferent fibers; **d.** afferent division, afferent fibers

2. **a.** dendrite; **b.** Nissl bodies; **c.** mitochondrion; **d.** nucleus; **e.** nucleolus; **f.** cell body; **g.** axon hillock; **h.** axolemma; **i.** axon; **j.** telodendrion; **k.** synaptic terminal

3. **a.** multipolar; **b.** unipolar; **c.** anaxonic; **d.** bipolar

Section 2 Review

1. **a.** action potential; **b.** electrical synapse; **c.** resting potential; **d.** gated channels; **e.** cholinergic synapses; **f.** hyperpolarization; **g.** local current; **h.** depolarization

2. **a.** acetylcholine (ACh); **b.** calcium ions (Ca^{2+}); **c.** synaptic vesicle; **d.** acetylcholinesterase (AChE); **e.** ACh receptor; **f.** sodium ions (Na^+); **g.** an action potential depolarizes the synaptic knob; **h.** calcium ions enter the cytoplasm of the synaptic knob; **i.** ACh is released through exocytosis; **j.** ACh binds to sodium channel receptors on the postsynaptic membrane, producing a graded depolarization; **k.** the depolarization ends as ACh is broken down into acetate and choline by AChE; **l.** the synaptic knob reabsorbs choline from the synaptic cleft and uses it to synthesize new molecules of ACh

3. In myelinated fibers, saltatory propagation transmits nerve impulses to the neuromuscular junctions rapidly enough to initiate muscle contractions and promote normal movements.

In axons that have become demyelinated, nerve impulses cannot be propagated, and so the muscles are not stimulated to contract. Eventually, the muscles atrophy because of the lack of stimulation (a condition termed disuse atrophy).

Chapter 12

Section 1 Review

1. a. white matter; b. dorsal root ganglion; c. lateral white column; d. posterior gray horn; e. lateral gray horn; f. anterior gray horn; g. posterior median sulcus; h. central canal; i. sensory nuclei; j. motor nuclei; k. anterior gray commissure; l. ventral root; m. anterior white commissure; n. anterior median fissure

2. a. anterior view; b. radial nerve; c. ulnar nerve; d. median nerve; e. posterior view

3. a. columns; b. conus medullaris; c. nerves; d. meninges; e. cauda equina; f. brachial plexus; g. dura mater; h. perineurium; i. gray ramus

Section 2 Review

1. a. divergence; b. convergence; c. serial processing; d. parallel processing; e. reverberation

2. a. receptor; b. sensory neuron; c. interneuron; d. spinal cord (CNS); e. motor neuron; f. effector

3. The withdrawal reflex illustrated in Question 2 is an innate, somatic, polysynaptic, spinal reflex.

4. a. ipsilateral reflex; b. withdrawal reflexes; c. gamma motor neuron; d. flexor reflex; e. visceral reflexes; f. acquired reflexes; g. contralateral reflex; h. reciprocal inhibition; i. reinforcement

Chapter 13

Section 1 Review

1. a. precentral gyrus; b. frontal lobe; c. lateral sulcus; d. temporal lobe; e. pons; f. central sulcus; g. postcentral gyrus; h. parietal lobe; i. occipital lobe; j. cerebellum; k. medulla oblongata

2. a. olfactory bulb (associated with cranial nerve I, olfactory), S;

b. oculomotor (III), M; c. trigeminal (V), B; d. facial (VII), B; e. glossopharyngeal (IX), B; f. vagus (X), B; g. optic (II), S; h. trochlear (IV), M; i. abducens (VI), M; j. vestibulocochlear (VIII), S; k. hypoglossal (XII), M; l. accessory (XI), M

3. a. thalamus; b. arcuate fibers; c. fornix; d. commissural fibers; e. basal nuclei

4. The sensory innervation of the nasal lining, or nasal mucosa, is by way of the maxillary branch of the trigeminal nerve (V). Irritation of the nasal lining increases the frequency of action potentials along the maxillary branch of the trigeminal nerve through the semilunar ganglion to reach centers in the midbrain, which in turn excite the neurons of the reticular activating system (RAS). Increased activity by the RAS can raise the cerebrum back to consciousness.

Section 2 Review

1. a. free nerve ending; b. root hair plexus; c. Merkel cells and tactile discs; d. tactile corpuscle; e. Ruffini corpuscle; f. lamellated corpuscle

2. a. lateral corticospinal tract of corticospinal pathway (conscious control of skeletal muscles throughout the body); b. rubrospinal tract of lateral pathway (subconscious regulation of muscle tone and movement of distal limb muscles); c. reticulospinal tract of medial pathway (subconscious regulation of muscle tone, and movements of the neck, trunk, and proximal limb muscles); d. vestibulospinal tract of medial pathway (subconscious regulation of muscle tone, and movements of the neck, trunk, and proximal limb muscles); e. tectospinal tract of medial pathway (subconscious regulation of muscle tone, and movements of the neck, trunk, and proximal limb muscles in response to bright lights, sudden movements, and loud noises); f. anterior corticospinal tract of corticospinal pathway (conscious control of skeletal muscles throughout the body); g. posterior column pathway (carries sensations of "fine" touch, pressure, vibration, and

proprioception); h. posterior spinocerebellar tract of spinocerebellar pathway (carries proprioceptive information about the position of skeletal muscles, tendons, and joints); i. lateral spinothalamic tract of spinothalamic pathway (carries pain and temperature sensations); j. anterior spinocerebellar tract of spinocerebellar pathway (carries proprioceptive information about the position of skeletal muscles, tendons, and joints); k. anterior spinothalamic tract of spinothalamic pathway (carries "crude" touch and pressure sensations)

3. a. anterior; b. posterior

4. Injuries to the motor cortex eliminate the ability to exert fine control over motor units, but gross movements may still be produced by cerebral nuclei using the reticulospinal or rubrospinal tracts.

Chapter 14

Section 1 Review

1. a. sympathetic division; b. thoracolumbar division; c. lumbar nerves; d. thoracic nerves; e. parasympathetic division; f. craniosacral division; g. cranial nerves III, VII, IX, X; h. sacral nerves; i. enteric nervous system

2. a. cervical sympathetic ganglia; b. sympathetic chain ganglia; c. coccygeal ganglia; d. sympathetic nerves; e. cardiac and pulmonary plexuses; f. celiac ganglion; g. superior mesenteric ganglion; h. splanchnic nerves; i. inferior mesenteric ganglion

3. a. splanchnic nerves; b. acetylcholine; c. parasympathetic activation; d. secrete norepinephrine; e. nicotinic, muscarinic; f. receptors; g. cholinergic; h. alpha, beta

Section 2 Review

1. a. limbic system and thalamus; b. hypothalamus; c. pons; d. spinal cord T_1–L_2; e. complex visceral reflexes; f. vasomotor; g. coughing; h. respiratory; i. sympathetic visceral reflexes; j. parasympathetic visceral reflexes

2. a. P; b. P; c. S; d. P; e. S; f. S; g. P; h. S; i. S; j. S; k. P; l. S; m. P

3. Even though most sympathetic postganglionic fibers are adrenergic, releasing norepinephrine, a few are cholinergic, releasing acetylcholine. This distribution of the cholinergic fibers via the sympathetic division provides a method of regulating sweat gland secretion and selectively controlling blood flow to skeletal muscles while reducing the flow to other tissues in a body wall to maintain homeostasis.

Chapter 15

Section 1 Review

1. a. G proteins b. olfaction c. stem cells d. odorant e. gustation f. Bowman glands g. cerebral cortex h. taste bud i. olfactory bulb j. olfactory cilia k. bitter l. depolarization m. lingual papillae

2. a. umami b. sour c. bitter d. salty e. sweet f. circumvallate papilla g. fungiform papilla h. filiform papilla

3. The olfactory sensory receptor cells are specialized neurons whose dendrites (also known as olfactory receptor cilia) contain receptor proteins. The binding of odorant molecules to the receptor proteins results in a depolarization of the receptor cell and the production of action potentials. In contrast, the membranes of the sensory receptor cells for taste, vision, equilibrium, and hearing are inexcitable and do not generate action potentials. These cells all form synapses with the processes of sensory neurons, which depolarize and produce action potentials when stimulated by chemical transmitters (neurotransmitters).

Section 2 Review

1. a. external ear b. middle ear c. inner ear d. auricle e. external acoustic meatus f. cartilage g. tympanic membrane h. auditory ossicles i. tympanic cavity j. petrous portion of temporal bone k. vestibulocochlear nerve (VIII) l. cochlea m. auditory tube

2. a. cerumen b. auricle c. endolymph d. tympanic cavity e. inner ear f. equilibrium g. tympanic membrane

h. stapes i. round window
j. incus k. ampulla l. temporal
lobe m. organ of Corti n. tectorial membrane o. decibel

3. The rapid descent in the elevator causes the otolith in the macula of the saccule of each vestibule to slide upward, producing the sensation of downward vertical motion. When the elevator abruptly stops, the otoliths do not. It takes a few seconds for them to come to rest in the normal position. As long as the otoliths are displaced, you will perceive movement.

Section 3 Review

1. a. posterior cavity b. choroid
c. fovea d. optic nerve e. optic disc f. retina g. sclera h. fornix
i. palpebral conjunctiva
j. ocular conjunctiva k. ciliary body l. iris m. lens n. cornea
o. suspensory ligaments p. ora serrata

2. a. rhodopsin b. palpebrae
c. crystallins d. sclera e. pupil
f. retina g. posterior cavity
h. rods i. posterior chamber
j. fovea k. occipital lobe
l. cones m. ganglion cells
n. vascular tunic o. optic disc

3. When light falls on the eye, it passes through the cornea and strikes the photoreceptors of the retina, bleaching (breaking down) many molecules of the pigment rhodopsin into retinal and opsin. After an intense exposure to light, a photoreceptor cannot respond to further stimulation until its rhodopsin molecules have been regenerated by the conversion of retinal molecules to their original shape and recombination with opsin molecules. The "ghost" image remains until the rhodopsin molecules are regenerated.

Chapter 16

Section 1 Review

1. a. catecholamines; b. thyroid hormones; c. tryptophan derivatives; d. peptide hormones; e. short polypeptides;
f. glycoproteins; g. small proteins; h. lipid derivatives;
i. eicosanoids; j. steroid hormones; k. transport proteins

2. a. thymus; b. pineal gland;
c. pancreatic islets; d. hypothalamus; e. kidneys; f. adrenal

glands; g. pituitary gland;
h. gonads; i. heart; j. digestive tract; k. thyroid gland; l. parathyroid glands

3. a. F cells; b. epinephrine;
c. direct communication;
d. tropic hormones; e. cyclic-AMP; f. secretes releasing hormones; g. androgens;
h. prostaglandins; i. FSH;
j. parathyroid glands

Section 2 Review

1. a. PTH and calcitonin; b. glucocorticoids; c. sympathetic activation; d. increase blood pressure and volume; e. protein synthesis; f. GH and glucocorticoids; g. gigantism; h. reduce blood pressure and volume;
i. homeostasis threat; j. PTH and calcitriol

2. a. release of natriuretic peptides; b. suppression of thirst;
c. Na$^+$ and H$_2$O loss from kidneys; d. reduced blood pressure; e. increased fluid loss; f. falling blood pressure and volume; g. erythropoietin released; h. renin released;
i. increased red blood cell production; j. aldosterone secreted; k. ADH secreted; l. rising blood pressure and volume

3. (1) The two hormones may have opposing or antagonistic effects, such as occurs between insulin (decreases blood glucose levels) and glucagon (increases blood glucose levels). (2) The two hormones may have an additive or synergistic effect, in which the net result is greater than the sum of each acting alone. An example is the enhanced glucose-sparing action of GH in the presence of glucocorticoids. (3) One hormone may have a permissive effect on another, in which the first hormone is needed for the second hormone to produce its effect. For example, epinephrine cannot alter the rate of tissue energy consumption without the presence of thyroid hormones. (4) The hormones may have integrative effects, in which the hormones may produce different but complementary results in specific tissues and organs. An example is the differing effects of calcitriol and parathyroid hormone (PTH) on tissues involved in calcium metabolism; calcitriol increases

calcium ion absorption from digestive system, and PTH inhibits osteoblast activity and enhances calcium ion reabsorption by the kidneys.

Chapter 17

Section 1 Review

1. a. plasma; b. water; c. solutes;
d. proteins; e. electrolytes, glucose, urea; f. albumins;
g. globulins; h. fibrinogen;
i. formed elements; j. erythrocytes; k. leukocytes;
l. platelets; m. neutrophils;
n. eosinophils; o. basophils;
p. lymphocytes; q. monocytes

2. a. red bone marrow; b. mature RBCs; c. matrix; d. monocytes; e. transport protein;
f. cross-reaction; g. lymphocytes; h. jaundice; i. venipuncture; j. pigment complex;
k. erythropoietin; l. platelets

3. During differentiation, the red blood cells of humans (and other mammals) lose most of their organelles, including nuclei and ribosomes. As a result, mature circulating RBCs can neither divide nor synthesize the structural proteins and enzymes required for cellular repairs.

Section 2 Review

1. a. common carotid; b. subclavian; c. brachiocephalic trunk; d. brachial; e. radial;
f. popliteal; g. fibular; h. aortic arch; i. celiac trunk; j. renal;
k. common iliac; l. external iliac; m. femoral; n. anterior tibial

2. a. vertebral; b. internal jugular;
c. brachiocephalic; d. axillary; e. cephalic; f. median antebrachial; g. ulnar; h. great saphenous; i. fibular; j. superior vena cava; k. inferior vena cava; l. internal iliac; m. deep femoral; n. posterior tibial

Chapter 18

Section 1 Review

1. a. aortic arch; b. superior vena cava; c. right pulmonary arteries; d. ascending aorta;
e. fossa ovalis; f. opening of coronary sinus; g. right atrium;
h. pectinate muscles; i. tricuspid valve cusp; j. chordae tendineae; k. papillary muscle;

l. right ventricle; m. inferior vena cava; n. pulmonary trunk;
o. pulmonary valve; p. left pulmonary arteries; q. left pulmonary veins; r. left atrium;
s. aortic valve; t. bicuspid valve cusp; u. left ventricle; v. interventricular septum; w. trabeculae carneae; x. moderator band

2. a. Deoxygenated blood flow: right atrium → right atrioventricular valve (tricuspid valve) → right ventricle → pulmonary semilunar valve. b. Oxygenated blood flow: left atrium → left atrioventricular valve (bicuspid valve) → left ventricle → aortic semilunar valve.

3. a. aorta; b. calcium ions;
c. myocardium; d. coronary sinus; e. fossa ovalis; f. tricuspid valve; g. anastomoses; h. intercalated discs; i. serous membrane;
j. endocardium; k. cardiac skeleton; l. aortic valve

Section 2 Review

1. a. left AV valve closes;
b. increasing, decreasing;
c. less than; d. aortic valve is forced open; e. aorta;
f. ventricular systole

2. a. "lubb" sound; b. automaticity;
c. P wave; d. stroke volume; e. cardiac output;
f. parasympathetic neurons;
g. bradycardia; h. sympathetic neurons; i. "dubb" sound;
j. tachycardia

Section 3 Review

1. The three primary factors influencing blood pressure and blood flow are cardiac output, blood volume, and peripheral resistance.

2. a. baroreceptors; b. medulla oblongata; c. venous return;
d. autoregulation; e. local vasodilators; f. chemoreceptors;
g. osmotic pressure; h. turbulence;
i. viscosity; j. net hydrostatic pressure; k. natriuretic peptides;
l. edema

3. a. arterioles; b. autonomic nervous system; c. increasing peripheral vasoconstriction;
d. increased vasodilation, increased venous return, increased cardiac output;
e. brain; f. nervous and endocrine

Chapter 19

Section 1 Review

1. a. tonsil; b. cervical lymph nodes;
c. right lymphatic duct;

d. thymus; e. cisterna chyli; f. lumbar lymph nodes; g. appendix; h. lymphatics of lower limb; i. lymphatics of upper limb; j. axillary lymph nodes; k. thoracic duct; l. lymphatics of mammary gland; m. spleen; n. mucosa-associated lymphoid tissue (MALT); o. pelvic lymph nodes; p. inguinal lymph nodes

2. a. lymphatic capillaries; b. thymic corpuscles; c. right subclavian vein; d. thoracic duct; e. lymphoid organs; f. lymph nodes; g. helper T cells and suppressor T cells; h. spleen; i. reticular epithelial cells; j. lymphopoiesis; k. cytotoxic T cells; l. tonsils; m. B cells; n. afferent lymphatics

Section 2 Review

1. a. physical barriers; b. phagocytes; c. immunological surveillance; d. interferons; e. complement; f. inflammatory response; g. fever

2. a. all of these; b. immunological surveillance; c. the complement system; d. interferons; e. pyrogens; f. phagocytes

3. The high body temperatures of a fever may inhibit some viruses and bacteria or speed their reproductive rates so that the disease runs its course more quickly. High body temperatures also accelerate the body's metabolic processes, which may help to mobilize tissue defenses and speed the repair process.

Section 3 Review

1. a. antibody; b. CD4 markers; c. acquired immunity; d. passive immunity; e. helper T cells; f. opsonization; g. Class I MHC; h. B lymphocytes; i. Class II MHC; j. IgM; k. costimulation; l. IgG; m. anaphylaxis

2. a. viruses; b. macrophages; c. natural killer (NK) cells; d. helper T cells; e. B cells; f. antibodies; g. cytotoxic T cells; h. suppressor T cells; i. memory T cells and B cells

Chapter 20

Section 1 Review

1. a. nasal cavity; b. hard palate; c. pharynx; d. glottis; e. trachea; f. right lung; g. external nares; h. larynx; i. primary bronchus; j. root of the lung; k. secondary bronchus; l. tertiary bronchus; m. bronchioles; n. terminal bronchiole; o. pulmonary lobule; p. alveolus

2. a. type II pneumocytes; b. trachea; c. type I pneumocytes; d. bronchodilation; e. terminal bronchiole; f. bronchoconstriction; g. respiratory bronchiole; h. respiratory membrane; i. phonation; j. pharynx; k. larynx; l. cystic fibrosis; m. respiratory mucosa; n. laryngeal prominence

Section 2 Review

1. a. inspiratory reserve volume (IRV): the amount of air that can be taken in above the resting tidal volume; b. resting tidal volume (V_T): the amount of air inhaled and exhaled during a single respiratory cycle while resting; c. expiratory reserve volume (ERV): the amount of air that can be expelled after a completely normal, quiet respiratory cycle; d. minimal volume: the amount of air remaining in the lungs if they were to collapse; e. inspiratory capacity: the amount of air that can be drawn into the lungs after completing a quiet respiratory cycle; f. total lung capacity: the total volume of the lungs; g. vital capacity: the maximum amount of air that can be moved into or out of the lungs in a single respiratory cycle; h. residual volume: the amount of air remaining in the lungs after a maximal exhalation; i. functional residual capacity (FRC): the amount of air that remains in the lungs after completing a quiet respiratory cycle

2. a. partial pressure; b. anoxia; c. Boyle's law; d. compliance; e. bicarbonate ion; f. lowers vital capacity; g. external intercostals; h. iron ion; i. hemoglobin releases more O_2; j. pneumotaxic centers; k. apneustic centers; l. hypocapnia; m. atelectasis; n. apnea

3. External respiration includes all the processes involved in the exchange of oxygen and carbon dioxide between the interstitial fluids and the external environment. Pulmonary ventilation, or breathing, is a process of external respiration that involves the physical movement of air into and out of the lungs. Internal respiration is the absorption of oxygen and the release of carbon dioxide by tissue cells.

Chapter 21

Section 1 Review

1. a. mesenteric artery and vein; b. mesentery; c. plica circulares; d. mucosa; e. submucosa; f. muscularis externa; g. serosa

2. a. esophagus; b. muscularis mucosae; c. lamina propria; d. plicae circulares; e. peristalsis; f. pacesetter cells; g. myenteric plexus; h. multi-unit smooth muscle cells; i. liver; j. segmentation; k. bolus; l. visceral smooth muscle cells; m. plasticity; n. sphincter

3. With a decrease in smooth muscle tone, general motility along the digestive tract decreases, and peristaltic contractions are weaker.

Section 2 Review

1. a. crown; b. neck; c. root; d. enamel; e. dentin; f. pulp cavity; g. gingiva; h. gingival sulcus; i. cementum; j. periodontal ligament; k. root canal; l. bone of alveolus

2. a. material in jejunum; b. gastrin; c. GIP; d. secretin and CCK; e. VIP; f. inhibits; g. acid production; h. insulin; i. bile; j. intestinal capillaries; k. gallbladder; l. nutrient utilization by tissues

3. Both parietal cells and chief cells are secretory cells found in the gastric glands of the wall of the stomach. However, parietal cells secrete intrinsic factor and hydrochloric acid, whereas chief cells secrete pepsinogen, an inactive proenzyme.

Section 3 Review

1. a. bile ductule; b. hepatocytes; c. central vein; d. interlobular septum; e. sinusoid; f. branch of hepatic artery; g. branch of hepatic portal vein; h. bile duct; i. portal area (portal triad)

2. a. starch; b. pancreas; c. common bile duct; d. hepatocytes; e. mumps; f. peptic ulcer; g. emulsification; h. pancreatic lipase; i. liver; j. lysozyme; k. submandibular glands; l. Kupffer cells; m. gallbladder; n. gallstones

3. Saliva (1) continuously flushes and cleans oral surfaces, (2) contains buffers that prevent the buildup of acids produced by bacterial action, and (3) contains antibodies (IgA) and lysozyme, which help control the growth of oral bacterial populations.

4. Such a blockage would interfere with the release of secretions into the duodenum by the pancreas, gallbladder, and liver. The pancreas normally secretes about 1 liter of pancreatic juice, a mixture of a variety of digestive enzymes and buffer solution. The blockage of pancreatic juice would lead to pancreatitis, an inflammation of the pancreas. Extensive damage to exocrine cells by the blocked digestive enzymes would lead to autolysis that could destroy the pancreas and result in the individual's death. Blockage of bile secretion from the common bile duct could lead to damage of the wall of the gallbladder by the formation of gallstones and to jaundice because bilirubin from the liver would not be excreted in the bile and, instead, would accumulate in body fluids.

Chapter 22

Section 1 Review

1. a. fatty acids; b. glucose; c. proteins; d. two-carbon chains; e. citric acid cycle; f. coenzymes; g. ATP; h. electron transport system; i. O_2; j. CO_2; k. H_2O

2. During fasting or starvation, other tissues shift to fatty acid or amino acid catabolism, conserving glucose for neural tissue.

3. a. citric acid; b. nutrient pool; c. anabolism; d. oxidative phosphorylation; e. cytochromes; f. coenzymes; g. oxygen; h. catabolism; i. nutrients abundant; j. ATP; k. citric acid cycle; l. water; m. nutrients scarce; n. acetate

Section 2 Review

1. a. insulin; b. skeletal muscle; c. B complex and C; d. ketone bodies; e. lipogenesis; f. urea formation; g. A, D, E, K; h. lipoproteins; i. deamination; j. lipolysis; k. uric acid; l. calorie; m. absorptive state; n. anorexia

2. a. jejunum; b. catabolized for energy; c. venous circulation via the thoracic duct; d. the excess amount is readily excreted in the urine

3. a. Essential amino acids are necessary in the diet because they cannot be synthesized by the body. The body can synthesize nonessential amino acids on demand.

b. (1) Proteins are difficult to break apart because of their complex three-dimensional structure. (2) The energy yield of proteins (4.32 Cal/g) is less than that of lipids (9.46 Cal/g). (3) The by-products of protein or amino acid catabolism are ammonium ions, a toxin that can damage cells. (4) Proteins form the most important structural and functional components of cells. Excessive protein catabolism would threaten homeostasis at the cellular to system levels of organization.

c. During the absorptive state, the intestinal mucosa is absorbing nutrients from the digested food. The focus of the postabsorptive state is the mobilization of energy reserves and the maintenance of normal blood glucose levels.

d. Liver cells can break down or synthesize most carbohydrates, lipids, and amino acids. The liver has an extensive blood supply and thus can easily monitor blood composition of these nutrients and regulate accordingly. The liver also stores energy in the form of glycogen.

4. It appears that Darla is suffering from ketoacidosis as a consequence of her anorexia. Because she is literally starving herself, her body is metabolizing large amounts of fatty acids and amino acids to provide energy and in the process is producing large quantities of ketone bodies (normal metabolites from these catabolic processes). One of the ketones that is formed is acetone, which can be eliminated through the lungs. This accounts for the smell of aromatic hydrocarbons on Darla's breath. The ketones are also converted into keto acids. In large amounts this lowers the body's pH and begins to exhaust the alkaline reserves of the buffer system. This is probably the cause of her arrhythmias.

Section 3 Review

1. **a.** sensible perspiration; **b.** leptin; **c.** thermoregulation; **d.** inhibits feeding center; **e.** nonshivering thermogenesis; **f.** 60 percent; **g.** peripheral vasoconstriction; **h.** neuropeptide Y; **i.** basal metabolic rate; **j.** ghrelin; **k.** 40 percent; **l.** peripheral vasodilation; **m.** insensible perspiration; **n.** shivering thermogenesis

2. **a.** all of these; **b.** radiation, conduction, convection, and evaporation; **c.** physiological responses and behavioral modifications; **d.** peripheral vasoconstriction; **e.** triglycerides in adipose tissue; **f.** low blood glucose levels

3. **a.** Because energy use at rest is powered by mitochondrial energy production, and mitochondrial energy production is proportional to oxygen consumption.

b. The heat-gain center functions in preventing hypothermia, or below-normal body temperature, by conserving body heat and increasing the rate of heat production by the body.

c. Nonshivering thermogenesis increases the metabolic rate of most tissues through the actions of two hormones, epinephrine and thyroid-stimulating hormone. In the short term, the heat-gain center stimulates the adrenal medullae to release epinephrine via the sympathetic division of the ANS. Epinephrine quickly increases the breakdown of glycogen (glycogenolysis) in liver and skeletal muscle, and the metabolic rate of most tissues. The long-term increase in metabolism occurs primarily in children as the heat-gain center adjusts the rate of thyrotropin-releasing hormone (TRH) release by the hypothalamus. When body temperature is low, additional TRH is released, which stimulates the release of thyroid-stimulating hormone (TSH) by the anterior lobe of the pituitary gland. The thyroid gland then increases its rate of thyroid hormone release, and these hormones increase rates of catabolism throughout the body.

Chapter 23

Section 1 Review

1. **a.** renal sinus: cavity within kidney that contains calyces, pelvis of the ureter, and segmental vessels; **b.** renal pelvis: funnel-shaped expansion of the superior portion of ureter; **c.** hilum: depression on the medial border of kidney, and site of the apex of the renal pelvis and the passage of segmental renal vessels and renal nerves; **d.** renal papilla: tip of the renal pyramid that projects into a minor calyx; **e.** ureter: tube that conducts urine from the renal pelvis to the urinary bladder; **f.** renal cortex: the outer portion of the kidney containing renal lobules, renal columns (extensions between the pyramids), renal corpuscles, and the proximal and distal convoluted tubules; **g.** renal medulla: the inner, darker portion of the kidney that contains the renal pyramids; **h.** renal pyramid: conical mass of the kidney projecting into the medullary region

containing part of the secreting tubules and collecting tubules; **i.** minor calyx: subdivision of major calices into which urine enters from the renal papillae; **j.** major calyx: primary subdivision of renal pelvis formed from the merging of four or five minor calyces; **k.** renal lobe: portion of kidney consisting of a renal pyramid and its associated cortical tissue; **l.** renal columns: cortical tissue separating renal pyramids; **m.** fibrous capsule (outer layer): covering of the kidney's outer surface and lining of the renal sinus

2. **a.** urinary bladder; **b.** nephrons; **c.** renal tubules; **d.** glomerulus; **e.** proximal convoluted tubule; **f.** papillary ducts; **g.** renal medulla; **h.** renal sinus; **i.** major calyces; **j.** ureter

Section 2 Review

1. **a.** nephron: functional unit of the kidney that filters and excretes waste materials from the blood and forms urine; **b.** proximal convoluted tubule: reabsorbs water, ions, and all organic nutrients; **c.** distal convoluted tubule: important site of active secretion; **d.** renal corpuscle: expanded chamber that encloses the glomerulus; **e.** nephron loop: portion of the nephron that produces the concentration gradient in the renal medulla; **f.** collecting system: series of tubes that carry tubular fluid away from the nephron; **g.** collecting duct: portion of collecting system that receives fluid from many nephrons and performs variable reabsorption of water and reabsorption or secretion of sodium, potassium, hydrogen, and bicarbonate ions; **h.** papillary duct: delivers urine to the minor calyx

2. **a.** renal corpuscle **b.** podocytes **c.** filtrate **d.** BCOP **e.** nephron loop **f.** aquaporins **g.** secretion **h.** aldosterone **i.** PCT **j.** ADH

3. The presence of plasma proteins and numerous WBCs in the urine indicates an increased permeability of the filtration membrane. This condition usually results from inflammation of the filtration membrane within the renal corpuscle. If the condition is temporary, it is probably an acute glomerular nephritis usually associated with a bacterial infection

(such as streptococcal sore throat). If the condition is long term, resulting in a nonfunctional kidney, it is referred to as chronic glomerular nephritis. The urine volume would be greater than normal because the plasma proteins increase the osmolarity of the filtrate.

Section 3 Review

1. **a.** internal urethral sphincter; **b.** trigone; **c.** external urethral sphincter; **d.** external urethral orifice; **e.** rugae; **f.** middle umbilical ligament; **g.** detrusor; **h.** transitional epithelium; **i.** urethra; **j.** micturition; **k.** stratified squamous epithelium

2. **a.** stretch receptors stimulated; **b.** afferent fibers carry information to sacral spinal cord; **c.** parasympathetic preganglionic fibers carry motor commands; **d.** detrusor muscle contraction stimulated; **e.** sensation relayed to thalamus; **f.** sensation of bladder fullness delivered to cerebral cortex; **g.** individual relaxes external urethral sphincter; **h.** internal urethral sphincter relaxes

3. Four primary signs and symptoms of urinary disorders are (1) changes in the volume of urine, (2) changes in the appearance of urine, (3) changes in the frequency of urination, and (4) pain.

4. cystitis/pyelonephritis: both conditions involve inflammation and infections of the urinary system, but cystitis refers to the urinary bladder, whereas pyelonephritis refers to the kidney; stress incontinence/overflow incontinence: both conditions involve an inability to control urination, but stress incontinence involves periodic involuntary leakage, whereas overflow incontinence involves a continual, slow trickle of urine; polyuria/proteinuria: both are abnormal urine conditions, but polyuria is the production of excessive amounts of urine, whereas proteinuria refers to the presence of protein in the urine

Chapter 24

Section 1 Review

1. **a.** osmoreceptors; **b.** fluid balance; **c.** plasma, interstitial fluid; **d.** sodium; **e.** ADH; **f.** hyponatremia; **g.** potassium; **h.** fluid

compartments; **i.** kidneys; **j.** fluid shift; **k.** hypertonic plasma; **l.** hypokalemia; **m.** dehydration; **n.** aldosterone

2. **a.** intracellular fluid (ICF); **b.** kidneys and sweat glands; **c.** from the cells into the ECF until osmotic equilibrium is restored; **d.** becomes hypotonic with respect to the ICF; **e.** all of these; **f.** out of the ICF and into the ECF

3. When tissues are burned, cells are destroyed and the contents of their cytoplasm leak into the interstitial fluid and then move into the plasma. Since potassium ions are normally found within cells, damage to a large number of cells releases relatively large amounts of potassium ions into the blood. The elevated potassium level would stimulate cells of the adrenal cortex to produce aldosterone, and the cells of the juxtaglomerular complex to produce renin. The renin would activate the angiotensin mechanism. Ultimately, angiotensin II would stimulate still more aldosterone secretion. The elevated levels of aldosterone would promote sodium retention and potassium secretion by the kidneys, thereby accounting for the elevated potassium levels in the patient's urine.

Section 2 Review

1. **a.** increased P_{CO_2}; **b.** acidosis; **c.** plasma pH decrease; **d.** increased; **e.** secreted; **f.** generated; **g.** decreased; **h.** decreased; **i.** increased; **j.** decreased P_{CO_2}; **k.** alkalosis; **l.** plasma pH increase; **m.** decreased; **n.** generated; **o.** secreted; **p.** increased; **q.** increased; **r.** decreased

2. The young boy has metabolic and respiratory acidosis. The metabolic acidosis resulted primarily from the large amounts of lactic acid generated by the boy's muscles as he struggled in the water. (The dissociation of lactic acid releases hydrogen ions and lactate ions.) Sustained hypoventilation during drowning contributed to both tissue hypoxia and respiratory acidosis. Respiratory acidosis developed as the P_{CO_2} increased in the ECF, increasing the production of carbonic acid and its dissociation into H^+ and HCO_3^-. Prompt emergency treatment is essential; the usual procedure involves some form of artificial or mechanical respiratory assistance (to increase the respiratory rate and decrease P_{CO_2} in the ECF) coupled with the intravenous infusion of a buffered isotonic solution that would absorb the hydrogen ions in the ECF and increase body fluid pH.

Chapter 25

Section 1 Review

1. **a.** prostatic urethra; **b.** ductus deferens; **c.** penile urethra; **d.** penis; **e.** epididymis; **f.** testis; **g.** external urethral orifice; **h.** scrotum; **i.** seminal gland; **j.** prostate gland; **k.** ejaculatory duct; **l.** bulbo-urethral gland

2. **a.** dartos muscle; **b.** spermatogonia; **c.** seminiferous tubules; **d.** interstitial cells; **e.** spermiogenesis; **f.** spermatogenesis; **g.** penis and scrotum; **h.** epididymis; **i.** nurse cells; **j.** semen; **k.** impotence; **l.** corpus spongiosum; **m.** luteinizing hormone (LH); **n.** follicle-stimulating hormone (FSH)

3. Normal levels of testosterone (a) promote the functional maturation of spermatozoa, (b) maintain the accessory organs of the male reproductive tract, (c) are responsible for the establishment and maintenance of male secondary sex characteristics, (d) stimulate bone and muscle growth, and (e) stimulate sexual behaviors and sexual drive (libido).

Section 2 Review

1. **a.** infundibulum; **b.** ovary; **c.** uterine tube; **d.** perimetrium; **e.** myometrium; **f.** endometrium; **g.** uterus; **h.** clitoris; **i.** labium minus; **j.** labium majus; **k.** fornix; **l.** cervix; **m.** external os; **n.** vagina

2. **a.** oocytes; **b.** menarche; **c.** vesicouterine pouch; **d.** broad ligament; **e.** rectouterine pouch; **f.** corpus luteum; **g.** uterine cycle; **h.** lactation; **i.** ovaries; **j.** LH surge; **k.** cervix; **l.** tubal ligation; **m.** GnRH; **n.** vulva

3. The endometrial cells have receptors for estrogens and progesterone and respond to these hormones as if the cells were still in the body of the uterus. Under the influence of estrogens, the endometrial cells proliferate at the beginning of the uterine (menstrual) cycle and begin to develop glands and blood vessels, which then further develop under the control of progesterone. This dramatic increase in tissue size exerts pressure on neighboring tissues or in some other way interferes with their function. It is the recurring expansion of tissue in an abnormal location that causes periodic pain.

Chapter 26

Section 1 Review

1. **a.** conception; **b.** neonate; **c.** amphimixis; **d.** gestation; **e.** hCG; **f.** relaxin; **g.** colostrum; **h.** amnion; **i.** blastocyst; **j.** chorion; **k.** inner cell mass; **l.** syncytial trophoblast; **m.** morula; **n.** embryonic disc

2. **a.** junction between the ampulla and isthmus of the uterine tube; **b.** ninth week after fertilization; **c.** second trimester; **d.** cleavage, implantation, placentation, embryogenesis; **e.** first trimester; **f.** respiratory, digestive, and urinary

3. It is very unlikely that the baby's condition is the result of a viral infection contracted during the third trimester. The development of organ systems occurs during the first trimester, and by the end of the second trimester, most organ systems are fully formed. During the third trimester, the fetus undergoes tremendous growth, but very little new organ formation occurs.

Section 2 Review

1. **a.** phenotype; **b.** alleles; **c.** homologous; **d.** karyotype; **e.** genotype; **f.** locus; **g.** heterozygous; **h.** genetics; **i.** homozygous; **j.** polygenic inheritance; **k.** simple inheritance; **l.** autosomes

2. **a.** Children who cannot roll the tongue must be homozygous recessive for the condition, and the only way a child can receive two recessive alleles from tongue-rolling parents is to have parents who are both heterozygous. **b.** The allele for achondroplasia dwarfism must be dominant. In order for two dwarf parents to produce a normal child, the child must be homozygous recessive, and thus both parents must be heterozygous. The probability that their next child (and all subsequent children) will be normal sized is 1 in 4, or 25 percent.

Illustration Credits

All illustrations by Medical & Scientific Illustration, except for the following by Imagineering: p. 7, p. 8, p. 12 bottom, p. 185 top, p. 207, p. 208, p. 213 top, p. 281 top, p. 283 center, p. 293 bottom, p. 294 top.

Photo Credits

Chapter 1 p. 9 top Andreas Reh/iStockphoto; p. 9 second Nick Byrne/iStockphoto; p. 13 top Andreas Vesalius. De humani corporis fabrica (Liber VII) from *Images from the History of Medicine* (NLM); p. 13 Custom Medical Stock Photo.

Chapter 2 p. 24 center iStockphoto.

Chapter 5 p. 64 top Biophoto Associates/Photo Researchers.

Chapter 6 pp. 67–68 left to right SIU BIOMED COM/Custom Medical Stock Photo; Dr. Kathleen Welch; Frederic H. Martini; Mark Nielsen (chezlark.com); Project Masters, Inc./The Bergman Collection; Scott Camazine/Photo Researchers; p. 71 bottom Gary Parker.com.

Chapter 7 pp. 73–74 all Ralph T. Hutchings; pp. 75–76 bottom all Ralph T. Hutchings; p. 79 center Ralph T. Hutchings; p. 80 bottom and 81 top Ralph T. Hutchings; p. 83 center Ralph T. Hutchings.

Chapter 8 p. 90 center Ralph T. Hutchings; p. 92 Mark D. Miller, S. Ward Casscells Professor of Orthopaedic Surgery, University of Virginia.

Chapter 9 p. 101 top Frederic H. Martini; p. 101 middle Terry Wilson/iStockphoto.

Chapter 13 p. 147 bottom Clinic/Getty Images.

Chapter 15 p. 168 top Custom Medical Stock Photo; p. 169 bottom Micheline Dubé/iStockphoto.

Chapter 16 p. 179 top Mikhail Kokhanchikov/iStockphoto; p. 180 top and bottom Sharon Dominick/iStockphoto; John Paul Kay/Peter Arnold/PhotoLibrary.

Chapter 17 p. 187 top Robert B. Tallitsch; p. 187 bottom Custom Medical Stock Photo; p. 188 top Shevelev Vladimir/Shutterstock.

Chapter 18 p. 201 center Ed Reschke/Peter Arnold Images/Photolibrary; p. 203 bottom William C. Ober.

Chapter 19 p. 219 bottom Biophoto Associates/Photo Researchers.

Chapter 20 p. 229 bottom Dana Spiropoulou/iStockphoto.

Chapter 22 p. 255 center APHP-PSL-GARO/PHANIE/Photo Researchers.

Chapter 23 p. 267 center Beranger/Photo Researchers; p. 267 bottom Photo Researchers.

Chapter 25 p. 286 top Brent Melton/iStockphoto.

Chapter 26 pp. 289–290 iStockphoto; p. 293 center Photo Lennart Nilsson/Albert Bonniers Forlag; p. 295 all Science Photo Library/Photo Researchers.